ELECTRONIC SYSTEM DESIGN: INTERFERENCE AND NOISE CONTROL TECHNIQUES

JOHN R. BARNES

Prentice-Hall, Inc., Englewood Cliffs, New Jersey 07632

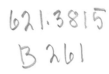
Library of Congress Cataloging-in-Publication Data

Barnes, John R. (date)
    Electronic system design.

    Bibliography: p.
    Includes index.
    1. Electronic circuits—Noise.   2. Electronic
circuit design.   3. Electromagnetic noise.
I. Title.
TK7867.5.B37 1987   621.3815'3   86-30229
ISBN 0-13-252123-7

Editorial/production supervision and
    interior design: *Gloria Jordan*
Cover design: *Wanda Lubelska*
Photo credit: *Teltronix, Inc.*
Manufacturing buyer: *S. Gordon Osbourne*

Printed in the United States of America

10 9 8 7 6 5 4 3 2 1

ISBN 0-13-252123-7   025

Prentice-Hall International (UK) Limited, *London*
Prentice-Hall of Australia Pty. Limited, *Sydney*
Prentice-Hall Canada Inc., *Toronto*
Prentice-Hall Hispanoamericana, S.A., *Mexico*
Prentice-Hall of India Private Limited, *New Delhi*
Prentice-Hall of Japan, Inc., *Tokyo*
Prentice-Hall of Southeast Asia Pte. Ltd., *Singapore*
Editora Prentice-Hall do Brasil, Ltda., *Rio de Janeiro*

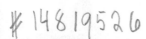

To Mike Kelly, Ted Hall, and Carolyn Herel,
for their patience, understanding, and support.

CONTENTS

TABLES

PREFACE

The greatest challenge in most electronic design projects is making the product work; meeting electrical noise limits (such as FCC Part 15, MIL-STD-461, and VDE 0871) often runs a close second. Unfortunately, our formal education usually neglects both of these subjects. Experienced colleagues may teach us some tricks of the trade, but for the most part we learn through trial and error.

This book is a practical guide to designing and building electronic equipment that (1) works, and (2) does not interfere with other equipment. It emphasizes design and construction practices that minimize electrical noise and electrical interference at little or no cost, thus saving development time and effort, reducing manufacturing costs, and providing more reliable products. This book should also help designers make better design decisions when they understand how noise problems occur and how they can be avoided.

Written mainly for electronic engineers and technicians, whose jobs and paychecks depend on their productivity, it will also benefit hobbyists and students by saving debug time and allowing them to complete more projects successfully.

The book contains noise-control techniques that apply from DC to hundreds of megahertz and to all types of electronic equipment. These techniques were gleaned from engineering books, technical reports, trade magazines, scholarly journals, product literature, and from extensive practical experience designing products and automated test equipment for industry.

The material is here presented with a maximum of application data and a minimum of theory. Scattered throughout the text are additional references, 16 tables of design data, 17 real-life noise/interference problems and their solutions, and nearly 300 design equations. These equations can be used to analyze noise/interference problems, and can be solved with a scientific calculator or incorporated into computer programs and spreadsheets. Most of the equations are accompanied by simple approximations, graphs, tables, or sample solutions.

Chapter 1 introduces the reader to the need for noise- and interference-control in electronic systems. Chapters 2 through 4 discuss important properties of components and wiring. Chapters 5 through 8 cover circuit-design techniques for analog, digital, interface, and power supply circuits. Chapters 9 through 13 cover physical-design techniques, including partitioning, grounding, bonding, circuit board layout, and cable design. Chapters 14 and 15 cover shielding and filtering—popular but expensive "band-aids" for noise problems. Chapter 16 discusses techniques for finding and fixing noise and interference problems. Nine appendices contain detailed information on worst-case design, electrical materials, wires, transmission lines, and electromagnetic waves. A glossary defines noise- and interference-control terms, and an annotated bibliography summarizes the contents of the best available books and technical reports on noise and interference-control techniques.

John R. Barnes
Lexington, Kentucky

ACKNOWLEDGEMENTS

Many people have contributed to my writing this book. I wish to give special thanks to the following people and organizations:

Lorayne Burns and Sharon Williams of the IBM Lexington (Kentucky) Technical Library helped with computer database searches and obtained many books, reports, and articles for me. The IBM Technical Libraries at East Fishkill, N.Y., Raleigh, N.C., Thornwood, N.Y., and Yorktown Heights, N.Y., also gave me access to their extensive collections.

The University of Kentucky libraries let me borrow many books and magazines from their collections. The libraries at California State University, Georgia Institute of Technology, North Carolina State University, Ohio State University, University of California, University of Louisville, University of Michigan, University of Tennessee, and Vanderbilt University allowed me to search through their collections and photocopy pertinent items.

Don Bush, group leader of the EMC Lab at IBM Lexington, and Clayton Paul, professor of Electrical Engineering at the University of Kentucky, shared their extensive experience, lent me documents from their personal collections, provided valuable guidance, and reviewed early drafts of this book. Ted Hall, manager of Test Engineering, provided the seed idea for this book in the spring of 1983.

Mike Kelly, manager of Card Test Engineering, has supported my research into noise-control techniques for the past three years and has read and critiqued every major draft of this book. Larry Collier, Steve Terrell, Robert Mauro, and Gadi Kaplan reviewed my original IBM technical report and the book's early drafts. John Humphreys, Mickey McInnes, and Bill Rowe also reviewed my original IBM technical report. Tim Krimm also read and critiqued the first draft of this book. Copy editor Bob Lentz did the excellent technical editing of the final manuscript.

My father, Robert S. Barnes, contributed the transmission-line terminating procedure described in Chapter 6. Brendan Mealy (Filtron Company), Claudette Lee (Ecos Electronics), Edward Siebenaler (Magnetics), J. Lee Plank (National Perforating Corporation), and Robert Bilby (Tecknit) helped me locate technical reports and technical information on some of the stickier problem areas. In addition, over 425 companies provided technical literature and information on their products.

And last but not least, I want to thank International Business Machines Corporation for letting me tackle this project, and for letting me use data from my 1984 IBM technical report *Electronic System Design*.

LIST OF SYMBOLS

UNITS

A   = amperes

dB = decibels, the ratio between power $P1$ and power $P2$ expressed as $10 \log_{10} (P1/P2)$

F   = farads

g   = grams (1 gram = 0.0022 pounds)

H   = henries

Hz = hertz

m   = meters (1 meter = 39.37 inches)

N   = newtons (1 newton = 0.2248 pounds force)

Pa = pascals (1 pascal = 1.45×10^{-4} PSI pressure)

s   = seconds

V   = volts

Ω = ohms

PREFIXES

G   = giga (10^9)

k   = kilo (10^3)

M   = mega (10^6)
m   = milli (10^{-3})
n   = nano (10^{-9})
p   = pico (10^{-12})
μ = micro (10^{-6})

COMPONENTS

C   = capacitor
CR = diode
L   = inductor
Q   = transistor
R   = resistor
S   = switch
T   = transformer
U   = integrated circuit

PHYSICAL VALUES

A = loop area (meter^2) or attenuation (dB)
A_a = attenuation of an aperture (dB)
B = inside diameter of a tube (meters) or re-reflection (dB)
B_a = re-reflection within an aperture (dB)
C = capacitance (farads)
c = speed of light in vacuum $\approx 2.99792548 \times 10^8$ meters/second
C_a = capacitance of an active transmission line (farads)
C_{in} = input capacitance of an integrated circuit (farads)
C_L = load capacitance (farads)
C_m = mutual capacitance between two conductors (farads)
C_{out} = output capacitance of an integrated circuit (farads)
C_p = parallel parasitic capacitance (farads)
C_q = capacitance of a quiet transmission line (farads)
C_u = capacitance per meter of transmission line (farads/meter)
D = outside diameter of a wire or tube (meters)
DF = dissipation factor (%)
ESL = equivalent series inductance of a component (henries)

ESR = equivalent series resistance of a component (ohms)

f = frequency (hertz)

f_b = breakpoint between low- and high-frequency effects (hertz)

f_c = self-resonant frequency for an RLC-network (hertz)

f_t = unity-gain frequency of a transistor (hertz)

GMD = geometric mean distance between current filaments (meters)

H = height (meters)

H_{fe} = common-emitter gain of a transistor

I = current (amperes)

I_a = current in active transmission line (amperes)

I_{arc} = minimum arcing current for contacts (amperes)

I_b = base current of a transistor (amperes)

I_c = collector current of a transistor (amperes)

I_d = drain current of a field-effect transistor (amperes)

I_{os} = short-circuit output current of an integrated circuit (amperes)

$K1$ = fraction open area of an array of apertures (dB)

$K2$ = penetration of the walls between the apertures (dB)

$K3$ = correction for closely spaced apertures (dB)

L = inductance (henries)

l = length (meters)

L_a = inductance of an active transmission line (henries)

l_c = critical length of a transmission line = $0.5 t_r / t_u$ (meters)

L_L = load inductance (henries)

L_m = mutual inductance between two conductors (henries)

L_q = inductance of a quiet transmission line (henries)

L_s = parasitic series inductance (henries)

L_u = inductance per meter of transmission line (henries/meter)

n = twists per meter for twisted pairs

NMH = noise margin in high state (volts)

NML = noise margin in low state (volts)

Q = quality factor of a capacitor or inductor

R = resistance (ohms) or reflection (dB)

r = distance from the source (meters)

R_a = equivalent shunt resistance of active circuit (ohms) or reflection from an aperture (dB)

R_{be}	=	base-emitter resistance of a Darlington transistor (ohms)
R_c	=	critical damping resistance for an *RLC*-circuit (ohms)
R_{ds}	=	drain-source resistance of a turned-on FET (ohms)
R_{fe}	=	far-end resistance of a quiet circuit (ohms)
R_{in}	=	input resistance of an integrated circuit (ohms)
R_L	=	load resistance (ohms)
R_{ne}	=	near-end resistance of a quiet circuit (ohms)
R_{outH}	=	output resistance in high state (ohms)
R_{outL}	=	output resistance in low state (ohms)
R_p	=	parallel parasitic resistance (ohms)
R_q	=	equivalent shunt resistance of quiet circuit (ohms)
R_S	=	output resistance of voltage source (ohms)
R_s	=	series parasitic resistance (ohms)
R_t	=	termination resistance (ohms)
S	=	spacing (meters)
SE	=	shielding effectiveness (dB)
sf	=	shape factor for transmission line
SRF	=	self-resonant frequency of a component (hertz)
T	=	thickness (meters)
t	=	time or pulse duration (seconds)
t_a	=	time constant of an active circuit (seconds)
t_f	=	fall time (seconds)
t_{HL}	=	high-to-low propagation delay (seconds)
t_{LH}	=	low-to-high propagation delay (seconds)
t_p	=	propagation delay (seconds)
t_q	=	time constant of a quiet circuit (seconds)
t_r	=	rise time (seconds)
t_u	=	propagation delay per meter of transmission line (seconds/meter)
V	=	voltage (volts)
V_a	=	voltage on active transmission line (volts)
V_{arc}	=	minimum arcing voltage for contacts (volts)
V_{bc}	=	backward crosstalk induced in a quiet line (volts)
V_{be}	=	base-emitter voltage of a transistor (volts)
V_{cc}	=	supply voltage for integrated circuits (volts)

V_{ce} = collector-emitter voltage of a transistor (volts)

V_d = voltage on a transmission line at the driver (volts)

V_f = voltage after a transition (volts) or diode forward voltage (volts)

V_{fc} = forward crosstalk induced in a quiet line (volts)

V_{fe} = voltage at the far end of a transmission line (volts)

V_i = voltage before a transition (volts)

V_{in} = input voltage to a circuit (volts)

V_m = voltage at the midpoint of a transmission line (volts)

V_{max} = rated working voltage for a capacitor

V_{ne} = voltage at the near end of a transmission line (volts)

V_{out} = output voltage of an integrated circuit (volts)

V_q = voltage on a quiet transmission line (volts)

V_r = voltage on a transmission line at the receiver (volts)

V_{ref} = reference voltage (volts)

V_S = output voltage of a voltage source (volts)

V_s = parasitic voltage induced on a wire (volts)

W = width (meters)

Z = impedance (ohms)

Z_a = impedance of an active transmission line (ohms)

Z_L = load impedance (ohms)

Z_m = mutual impedance between two conductors (ohms)

Zo = nominal impedance of a transmission line (ohms)

Zo' = impedance of a terminated transmission line (ohms)

Z_q = impedance of a quiet transmission line (ohms)

Z_s = shield impedance (ohms)

Z_w = wave impedance (ohms)

α = attenuation constant for a material

β = phase constant for a material

ΔV = voltage step (volts)

ΔV_{out} = output voltage swing (volts)

δ = skindepth = $[\rho/(\pi\mu_v\mu_r f)]^{1/2}$ meters

ϵ = permittivity (farads/meter)

ϵ_r = relative permittivity (1 for vacuum, $\approx$1 for air and good conductors, sometimes called K, or dielectric constant)

$\epsilon_{r'}$ = average relative permittivity around a transmission line

ϵ_v = permittivity of vacuum $\approx 8.85418782 \times 10^{-12}$ farads/meter

η = intrinsic impedance of a material.

Θ = angle of twist of twisted pair (degrees)

θ = angle with respect to z-axis (degrees)

λ = wavelength = $c/f \approx 2.99792548 \times 10^8/f$ meters in air or vacuum

μ = permeability (henries/meter)

μ_r = relative permeability (≈ 1 for insulators and nonmagnetic conductors)

$\mu_{r'}$ = average relative permeability around a transmission line

μ_v = permeability of vacuum $\approx 1.25663706 \times 10^{-6}$ henries/meter

ρ = resistivity (ohm-meters)

σ = intrinsic propagation constant of a material

ϕ = angle with respect to x-axis (degrees)

MISCELLANEOUS SYMBOLS

GND   = ground

k = ratio of aperture impedance to wave impedance

j = $(-1)^{1/2}$

π ≈ 3.14159265

1

INTRODUCTION

Every electrical engineer and technician enjoys designing and building equipment that costs less than expected, is finished ahead of schedule, operates reliably, and doesn't interfere with other equipment. But all too often we miss several of these goals because of electrical noise and electrical interference problems. If we can't solve these problems in a reasonable amount of time, we may have to abandon the project or start over from scratch—wasting all the time, money, and effort that we had invested in the project. For example, if a digital system exceeds Federal Communications Commission (FCC) limits on conducted and radiated emissions (FCC part 15 subpart J), we are not allowed to sell the system *or even offer to sell the system* in the United States. Most European countries have similar requirements, Germany's VDE 0871 being the one of the toughest, and most electronic equipment for the United States military must meet MIL-STD-461.

Most electrical engineering classes and books either ignore electrical noise or limit the discussion to thermal noise. As a result most engineers' first *real* encounter with electrical noise is when they are trying to debug their first system. This lack of preparation often has three serious side-effects: debug takes much longer than expected, the designers become very frustrated, and the noise-suppression components needed to solve the problem add 10 to 15 percent to the manufacturing cost!

Early in my career I discovered a much better method—*design noise control into the product from the beginning*! This is a four-step process: (1) Understand what types of noise problems can occur. (2) Design the circuitry to reduce or eliminate as many of these problems as possible. (3) Design circuit boards, cables, and the physical structure to eliminate as many noise problems as possible, and to allow for the easy addition of noise-suppression components if necessary. (4) Debug the system one piece at a time, making sure that each subsystem is assembled properly, operates properly, and doesn't have any noise problems before going on to the next. By designing the system right the first time I usually finish ahead of schedule, under the estimated cost, and feel good about my work.

This book contains the lessons that I have learned from fourteen years extensive (and sometimes expensive) experience as a practicing electronics engineer, from discussions with co-workers, and from three years of research into methods for controlling electrical noise. Chapters 2 through 4 cover noise generators, noise receivers, and noise-coupling paths. Chapters 5 through 8 cover electrical design techniques to minimize noise problems. Chapters 9 through 15 cover physical design techniques to control noise coupling. Chapter 16 covers techniques for finding, identifying, and fixing noise problems. The appendices contain detailed information on analyzing electrical noise problems and comparing design options.

As part of the research for this book, I searched the collections of eighteen large technical libraries across the country. In all, I examined well over eleven kilometers of books, magazines, and conference proceedings. In this mass of material I found 180 books, 73 technical reports, and about 2300 magazine articles and technical papers with information on controlling electrical noise.

Of these, seven sources stand out above all the rest. Henry W. Ott's *Noise Reduction Techniques in Electronic Systems* covers design techniques for avoiding electrical noise problems and was the best single source that I found. William R. Blood's *MECL System Design Handbook* covers the design of high-speed systems in very fine detail. R. Kenneth Keenan's *Digital Design for Interference Specifications* covers noise-control techniques for digital logic. Donald R. J. White's *A Handbook Series on Electromagnetic Interference and Compatibility, Volume 3*, covers techniques for fixing noise problems in existing systems. A technical report by Filtron Company, *Interference Reduction Guide for Design Engineers, Volume 1* (available through NTIS in Springfield, VA—AD 619 666), has a wealth of data on design for radio-frequency interference (RFI) control, including many graphs and tables of design data. And the two outstanding magazines in the field are *EMC Technology* and the *IEEE Transactions on Electromagnetic Compatibility*.

2

PASSIVE COMPONENTS

Most electronics books assume that resistors, capacitors, and inductors are linear and well behaved, having impedances

$$Z = \frac{V}{I} = R \text{ ohms for } R\text{-ohm resistors,}$$

$$Z = \frac{V}{I} = \frac{1}{j2\pi fC} \text{ ohms for } C\text{-farad capacitors, and}$$

$$Z = \frac{V}{I} = j2\pi fL \text{ ohms for } L\text{-henry inductors}$$

at f hertz (Figure 2-1), where Z, V, and I are vector quantities (Appendix I). In reality, however, all components have parasitic resistance, parasitic capacitance, and parasitic inductance. These parasitics usually

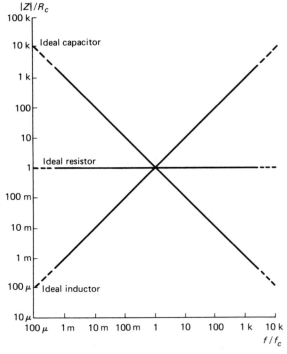

Figure 2-1 Impedance of ideal components

7

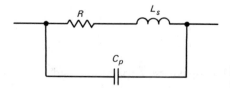

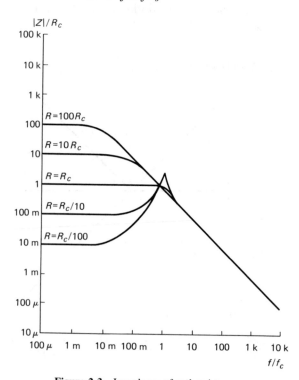

Figure 2-2 Lumped-impedance model for real resistors

have insignificant effects at low frequencies but tend to dominate component behavior at high frequencies.

Figure 2-2 shows the lumped-impedance model for real resistors. R is the desired resistance in ohms, L_s is the parasitic series inductance in henries, and C_p is the parasitic parallel capacitance in farads, due to the resistor's leads and internal construction. At f hertz the resistor's impedance is

$$Z = \frac{V}{I} = \frac{1}{\dfrac{1}{R + j2\pi f L_s} + j2\pi f C_p} \text{ ohms.}$$

Figure 2-3 Impedance of real resistors

Figure 2-3 shows some typical impedance-vs.-frequency curves for real resistors. Notice two distinct behaviors: the impedance of high-value resistors starts level and then drops, while the impedance of low-value resistors starts level, rises to a peak, and then drops.

If we try various values for R, L_s, and C_p, we find that $R \approx 1.55(L_s/C_p)^{1/2}$ ohms is the lowest resistance that doesn't have a peak in its impedance curve. So let's define

$$R_c = 1.55 \left(\frac{L_s}{C_p} \right)^{1/2} \text{ ohms}$$

as the *critical resistance* for a resistor. If a resistor has resistance $R \geqslant R_c$ ohms, we can approximate its impedance by

$$|Z| \approx R \text{ ohms for } f \leqslant \frac{1}{2\pi RC_p} \text{ hertz}$$

and

$$|Z| \approx \frac{1}{2\pi f C_p} \text{ ohms for } f > \frac{1}{2\pi RC_p} \text{ hertz.}$$

If a resistor has resistance $R < R_c$ ohms, then L_s and C_p can resonate at $f_c = 1/[2\pi(L_s C_p)^{1/2}]$ hertz. In this case we can approximate the resistor's impedance by

$$|Z| \approx R \text{ ohms for } f < \frac{R}{2\pi L_s} \text{ hertz,}$$

$$|Z| \approx 2\pi f L_s \text{ ohms for } \frac{R}{2\pi L_s} \leqslant f < \frac{f_c}{3} \text{ hertz,}$$

rising to

$$|Z| = \left[\left(\frac{L_s}{RC_p} \right)^2 + \frac{L_s}{C_p} \right]^{1/2} \text{ ohms for } f = f_c \text{ hertz,}$$

then falling to

$$|Z| \approx \frac{1}{2\pi f C_p} \text{ ohms for } f > 3f_c \text{ hertz.}$$

Table 2-1 shows the range of parasitic inductances, parasitic capacitances, and resonant frequencies exhibited by common types of resistors. In general, we want the resonant frequency of a resistor to be

TABLE 2-1   **Characteristics of Common Resistors**

Resistor Type	L_s (nH)	C_p (pF)	f_c (MHz)
Bulk metal	3–100	0.1–1.0	500–3000
Carbon composition	5–30	0.1–1.5	750–2000
Carbon film	15–700	0.1–0.8	300–1500
Metal film	15–700	0.1–0.8	300–1500
Surface mount	0.2–3	0.01–0.08	500–4000
Wirewound	47–25,000	2–14	8–200
Wirewound (noninductive)	2–600	0.1–5	90–1500

much higher than the circuit's operating frequency in order to avoid drastic impedance shifts.

Figure 2-4 shows the lumped-impedance model for real capacitors. C is the desired capacitance in farads. L_s is parasitic series inductance in henries, R_s is series resistance in ohms, and R_p is leakage resistance in ohms, all due to the capacitor's leads and internal construction. At f hertz the capacitor's impedance is

$$Z = \frac{V}{I} = \frac{1}{j2\pi f C + \dfrac{1}{R_p}} + j2\pi f L_s + R_s \text{ ohms.}$$

Figure 2-5 shows some typical impedance-vs.-frequency curves for real capacitors. If the capacitor has high series resistance, the impedance curve flattens out near the self-resonant frequency $f_c = 1/[2\pi(CL_s)^{1/2}]$ hertz. If the capacitor has low series resistance, the impedance curve takes a sharp dip near f_c.

If we play around with the equation for a capacitor's impedance, we discover that $R_s \approx 1.41 \, (L_s/C)^{1/2}$ ohms provides the shortest, smoothest transition between capacitive behavior ($f < f_c$) and inductive behavior ($f > f_c$). So let's define

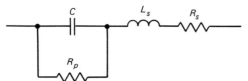

Figure 2-4 Lumped-impedance model for real capacitors

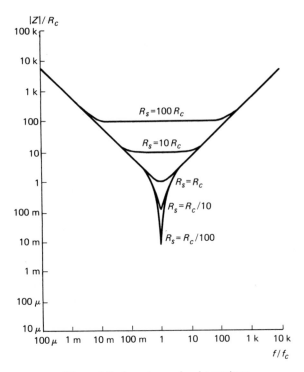

Figure 2-5 Impedance of real capacitors

$$R_c = 1.41 \left(\frac{L_s}{C_p}\right)^{1/2} \text{ ohms}$$

as the *critical series resistance* for a capacitor. If a capacitor has series resistance $R_s \geqslant R_c$, we can approximate its impedance by

$$|Z| \approx \frac{1}{2\pi f C} \text{ ohms for } f < \frac{1}{2\pi R_s C} \text{ hertz,}$$

$$|Z| \approx R_s \text{ ohms for } \frac{1}{2\pi R_s C} \leqslant f \leqslant \frac{R_s}{2\pi L_s} \text{ hertz,}$$

and

$$|Z| \approx 2\pi f L_s \text{ ohms for } f > \frac{R_s}{2\pi L_s} \text{ hertz.}$$

If $R_s < R_c$, C and L_s can resonate near f_c. In this case we can approximate the impedance by

$$|Z| \approx \frac{1}{2\pi f C} \text{ ohms for } f < \frac{f_c}{3} \text{ hertz,}$$

falling to

$$|Z| = R_s \text{ ohms for } f = f_c \text{ hertz,}$$

and then rising to

$$|Z| \approx 2\pi f L_s \text{ ohms for } f > 3f_c \text{ hertz.}$$

Table 2-2 shows the range of series inductances, series resistances, leakage resistances, and self-resonant frequencies exhibited by common types of capacitors. We usually want a capacitor's self-resonant frequency to be much higher than the circuit's operating frequency. This can be a problem with high-value capacitors. One simple solution is to put small high-quality capacitors in parallel with the large capacitors (see Chapters

TABLE 2-2   Characteristics of Common Capacitors

Capacitor Type	L_s (nH)	R_s (Ω)	R_p	f_c (MHz)
Aluminum, 2-lead	2–100	0.003–100	≥ 17 ΩF/C	0.001–0.5
Aluminum, 4-lead	0.04–2	0.011–2.6	≥ 35 ΩF/C	0.02–1
Aluminum, stacked-foil	1–2	0.001–0.3	≥ 35 ΩF/C	0.02–1
Ceramic, axial/disc	1–30	0.005–27	$\geq 5,000$ MΩ	2–800
Ceramic, feedthrough	0,001–1	0.6–360	$\geq 1,000$ ΩF/C	160–10,000
Ceramic, surface-mount	0.06–30	0.005–5	$\geq 1,000$ ΩF/C	2–60,000
Glass	1.4–10	0.01–2	$\geq 10,000$ MΩ	6–1,000
Mica	0.52–25	0.1–47	≥ 700 MΩ	5–7,000
Mylar	5–50	0.01–5	$\geq 1,000$ ΩF/C	2–35
Paper	6–160	1–16	≥ 20 ΩF/C	2–15
Polycarbonate	12–55	0.001–5	$\geq 15,000$ ΩF/C	0.1–15
Polyester	5–50	0.01–5	$\geq 1,000$ ΩF/C	2–35
Polypropylene	6–75	0.001–0.5	$\geq 30,000$ ΩF/C	0.3–15
Polystyrene	8–50	0.16–3.2	$\geq 90,000$ MΩ	5–100
Porcelain	0.02–2	0.01–0.8	$\geq 10,000$ MΩ	35–16,000
Stacked-film	2–10	0.5–1.3	$\geq 1,000$ ΩF/C	1–80
Tantalum, feedthrough	4–20	0.7–20	≥ 50 ΩF/C	25–1,000
Tantalum, foil	18–50	0.05–0.5	≥ 50 ΩF/C	0.02–1
Tantalum, solid	0.6–20	0.1–10	≥ 50 ΩF/C	0.3–50
Tantalum, surface-mount	0.02–1.5	0.04–3	≥ 50 ΩF/C	1–20
Tantalum, wet	2.3–50	0.05–15	≥ 160 ΩF/C	0.02–1
Teflon	15–55	0.02–1	$\geq 90,000$ ΩF/C	0.7–10

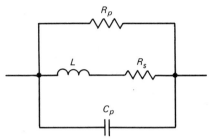

Figure 2-6 Lumped-impedance model for real inductors

6 and 8 for some examples). This technique also helps compensate for the normal increase in series resistance as electrolytic capacitors age and thus maintains the bypassing effectiveness of the circuit. To filter out very-high-frequency noise we may need to use feedthrough capacitors, mounted in shields to provide the necessary input-output isolation.

Figure 2-6 shows the lumped-impedance model for real inductors. L is the desired inductance in henries. R_p is leakage resistance, plus any core losses, in ohms. R_s is the winding resistance in ohms, and C_p is parasitic capacitance in farads, due to the inductor's leads and internal construction. (*Note:* Unshielded open-core inductors make excellent antennas for magnetic fields.) At f hertz the inductor's impedance is

$$Z = \frac{V}{I} = \frac{1}{\dfrac{1}{j2\pi fL + R_s} + \dfrac{1}{R_p} + j2\pi fC_p} \text{ ohms.}$$

Figure 2-7 shows some typical impedance-vs.-frequency curves for real inductors. Notice the strong resemblance to the impedance curves for low-value resistors in Figure 2-3. If R_p is very large and R_s is very small, we can use the following approximation for the inductor's impedance. Letting $f_c = \dfrac{1}{2\pi (LC_p)^{1/2}}$ hertz,

$$|Z| \approx R_s \text{ ohms for } f < \frac{R_s}{2\pi L} \text{ hertz,}$$

$$|Z| \approx 2\pi fL \text{ ohms for } \frac{R_s}{2\pi L} \leq f < \frac{f_c}{3} \text{ hertz,}$$

rising to

$$|Z| = \left[\left(\frac{L}{R_s C_p}\right)^2 + \frac{L}{C_p} \right]^{1/2} \text{ ohms for } f = f_c \text{ hertz,}$$

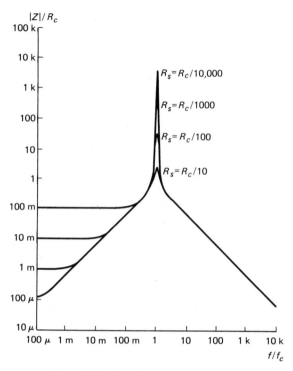

Figure 2-7   Impedance of real inductors

then falling to

$$|Z| \approx \frac{1}{2\pi f C_p} \text{ ohms for } f > 3f_c \text{ hertz.}$$

Common radio-frequency (RF) chokes have $0.2 \ \Omega < R_s < 5 \ \Omega$ and $1.5 \text{ pF} < C_p < 4 \text{ pF}$. Most surface-mount inductors have $R_s < 10 \ \Omega$ and $0.2 \text{ pF} < C_p < 20 \text{ pF}$.

Figure 2-8 shows the lumped-impedance model for real transformers. One winding has inductance $L1$ (henries), resistance R_{s1} (ohms), and parasitic capacitance C_{p1} (farads). The other winding has inductance $L2$, resistance R_{s2}, and parasitic capacitance C_{p2}. The windings are linked by mutual inductance L_m and interwinding capacitance C_m. Figure 2-9 shows the types of transients produced by a transformer, many of them due to the interwinding capacitance. Standard transformers have 10 pF $< C_m < 50 \text{ pF}$, while split-bobbin transformers (Figure 2-10(a)) reduce C_m to $\approx 5 \text{ pF}$. Faraday-shielded transformers, with electrostatic shields around the windings, can reduce C_m to $\approx 0.001 \text{ pF}$ with proper lead dress and shield connections (see Chapter 8).

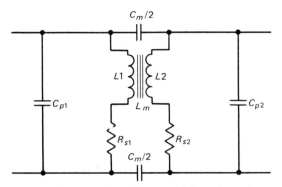

Figure 2-8   Lumped-impedance model for real transformers

Another problem area is the transformer's core. A stacked-lamination core has substantial magnetic-field leakage, which can couple noise into wiring and vacuum tubes. One way to reduce these fields is to wrap copper or aluminum tape around the transformer to form a shorting ring (Figure 2-10(c)). For maximum efficiency, transformers are often operated near saturation. If the core becomes saturated, however, the transformer will generate voltage spikes and harmonics of the input voltage. To eliminate these problems many engineers select toroidal transformers with high-permeability cores.

Toroidal transformers, however, are not a cure-all. Because of their high magnetic efficiency, toroidal transformers also tend to draw high inrush currents when power is first applied—up to 15 times the steady-state input current! To limit this inrush current we need to keep the core from saturating. One method is to cut the core in half and glue it back together again, adding two small air gaps to the magnetic circuit. The main drawback to this method is the slight increase in core losses. To hold core losses down while protecting the transformer from saturation, several companies offer two-part cores that combine an ungapped core with a gapped core. Another possibility is the quadracoil transformer (Figure 2-10(b)), which has a closed magnetic loop for low magnetic-field leakage, split bobbins for low interwinding capacitance, and high leakage inductance for transient suppression and noise filtering.

When selecting the components to be used in a circuit we should be guided by our previous experience, measurements on sample components, component specifications, and the component data in Tables 2-1 and 2-2. These tables list L_s, C_p, R_s, R_p, and f_c for various types of resistors and capacitors, gleaned from the product literature of nearly 200 companies. The tables show the range of values found in the datasheets

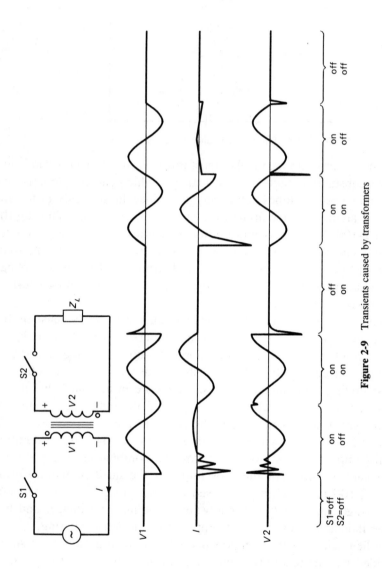

Figure 2-9 Transients caused by transformers

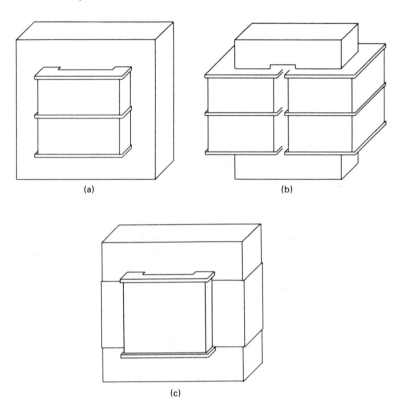

(a)　　　　　　　　　　　　　　(b)

(c)

Figure 2-10　Methods for reducing transformer noise: (a) split-bobbin, (b) quadra-coil, (c) shorting-ring

and do not represent any specific vendor's capabilities. We can use this data to sort components into three groups: (1) obviously unsuited for our application, (2) need data on specific parts, and (3) almost guaranteed to work in our system.

For components in the second group, we need to get datasheets from the manufacturers. If we are lucky, the datasheets will have the information we are looking for. Datasheets for inductors usually list the self-resonant frequency ($SRF = f_c$) and the maximum DC resistance ($DCR = R_s$), and we can usually ignore R_p. Given L and f_c, we can compute C_p.

For capacitors we usually aren't as lucky. If the datasheet has a graph showing impedance-vs.-frequency, the center of the dip is f_c hertz and the lowest impedance is R_s ohms. Given C and f_c, we can compute L_s. If the datasheet lists the dissipation factor (DF) or the quality factor (Q), we can compute R_s using the equation $DF = 1/Q = 2\pi f R_s C$. The

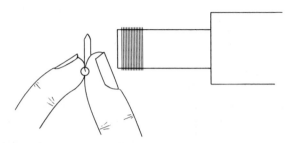

Figure 2-11 Measuring self-resonant frequency with a grid-dip meter

minimum leakage resistance (R_p or R_pC) or the maximum leakage current ($I = V_{max}/R_p$) may also be specified.

A few datasheets for resistors list C_p or show impedance-vs.-frequency graphs from which we can estimate C_p. For nonwirewound resistors, L_s is approximately equal to the inductance of a piece of the lead wire as long as the body of the resistor (see Appendix D).

If we have sample parts, we can measure them with an impedance bridge (*LRC* meter) and a grid-dip (gate-dip) meter by:

1. measuring the component value with the impedance bridge;
2. cutting the leads to the required length;
3. bending the leads in a circle, and soldering their ends together;
4. holding the component about 5 mm in front of the grid-dip meter's coil (Figure 2-11);
5. finding the frequency that minimizes the grid current—this is f_c.

From f_c and the component value we can compute the unknown value.

RECOMMENDED READING

BOWICK, CHRIS, *RF Circuit Design*. Indianapolis: Howard W. Sams & Co., Inc., 1982.

FICCHI, ROCCO F., ed., *Practical Design for Electromagnetic Compatibility*. New York: Hayden Book Co., Inc., 1971.

GREENWOOD, ALLAN, *Electrical Transients in Power Systems*. New York: Wiley-Interscience, 1971.

WOODY, JIMMY A., and CARMEN A. PALUDI, JR., "Modeling Techniques for Discrete Passive Components to Include Parasitic Effects in EMC Analysis and Design," *1980 IEEE International Symposium on Electromagnetic Compatibility*, Baltimore, 39–45.

3

ACTIVE
COMPONENTS

Active devices act as noise sources, noise-coupling paths, and noise receivers. Many active devices are designed to generate signals with sharp rising and falling edges, but these rapid transitions tend to cause noise problems in the 10- to 300-MHz band. Some active devices may oscillate at 0.1 to 20 MHz because of parasitic capacitance between their input pins and output pins; parasitic capacitance can also couple noise between signals that are supposed to be isolated from one another. Finally, non-linear devices can rectify high-frequency signals, generating harmonics and other spurious signals.

The simplest active devices are diodes and rectifiers, having single PN junctions. When a diode is reverse-biased, its junction is depleted of charge, but parasitic capacitance can still couple high-frequency noise through the device. If we suddenly forward-bias the diode, it will have high impedance for a few nanoseconds, producing a small forward-transient voltage spike (V in Figure 3-1). Then, if we suddenly reverse-bias the diode again, we must remove the charge in the junction to stop the current flow. For a fraction of a second the diode looks like a dead short, producing a large reverse-recovery current spike (I in Figure 3-1). For example, if an ordinary rectifier is carrying 0.5 A, on turn-off the reverse-recovery spike may reach 1.75 A and last for 40 μs. To reduce these spikes we can use high-current high-voltage diodes or fast-recovery diodes. (*Note:* Schottky hot-carrier diodes do not store charge like ordinary diodes, but their high parasitic capacitance produces similar effects.)

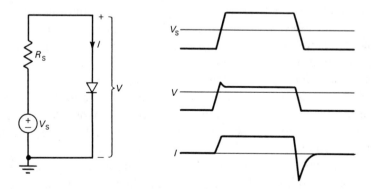

Figure 3-1 Effects of charge storage on diode turn-on/turn-off

Zener diodes usually have 10–7000 pF parasitic capacitance, which can keep them from clamping very fast spikes. Most zeners also have negative-resistance regions near their breakdown knee and can generate 1–1000 μV of white noise there. This noise is especially serious in alloy zeners. We can reduce zener noise by using diffused zeners instead of alloy zeners, by increasing the operating current to get well above the knee, and by bypassing the zeners with small ceramic capacitors.

Silicon-controlled rectifiers (SCRs) and triacs can put large spikes on AC lines, because they turn on quickly and usually switch large currents. An SCR or triac can be turned on accidentally by a spike at its anode, coupling into the gate through internal parasitic capacitance (the dV/dt problem). Zero-crossing detectors (to trigger SCRs and triacs at minimum load current) and RC-snubber networks (to slow the current rise at the anode) can reduce these problems.

The base-emitter junctions of bipolar transistors exhibit all the characteristics and problems of PN diodes. In addition, parasitic capacitance can couple high-frequency noise between the base, emitter, and collector and may make the transistor oscillate—very common when a transistor with unity-gain frequency $f_t \geq 100$ MHz is operated at less than $0.2f_t$. In general we should use the slowest transistors that will work in the circuit. A final problem is that the base-emitter and base-collector junctions can rectify high-frequency noise, upsetting the transistor's biasing. This can reduce the transistor's gain, or turn on transistors that were supposed to be turned off!

Parasitic capacitance can couple noise between the gate, drain, and source of a field-effect transistor (FET). In high-impedance circuits these feedthrough spikes may exceed the desired signals. Parasitic capacitance can also make FETs oscillate. A small gate resistor (100 Ω to 2 kΩ), or a ferrite bead on the gate lead, can help prevent these spurious oscillations.

Vacuum tubes suffer from a number of noise problems. They can oscillate, pick up stray electromagnetic fields, develop microphonics from shock or vibration, pick up hum from AC heaters, and develop leakage between the cathode, grid, plate, and other elements.

Parasitic capacitance slows down operational amplifiers (opamps) and limits their slew rate, which can make them oscillate or go into saturation. Most opamps will oscillate at 0.5–4 MHz if they are forced to drive reactive outputs.

Digital integrated circuits (ICs) generate trapezoidal pulses with very fast rising and falling edges (Figure 3-2). At f hertz these pulses have voltage

$$V = 2V_S t \frac{\sin{(\pi f t)}}{\pi f t} \frac{\sin\left(2\pi f \dfrac{t_r t_f}{t_r + t_f}\right)}{2\pi f \dfrac{t_r t_f}{t_r + t_f}} \text{ volts.}$$

The positions of the peaks and valleys of this equation are very sensitive to the exact values of t, t_r, and t_f. To be safe, we should use the upper bound of this equation:

$$V \leqslant 2V_S t \text{ volts for } f \leqslant f_1 = \frac{1}{\pi t} \text{ hertz,}$$

$$V \leqslant \frac{2V_S}{\pi f} \text{ volts for } f_1 < f \leqslant f_2 = \frac{t_r + t_f}{2\pi t_r t_f} \text{ hertz,}$$

and

$$V = \frac{V_S (t_r + t_f)}{\pi^2 f^2 t_r t_f} \text{ volts for } f > f_2 \text{ hertz.}$$

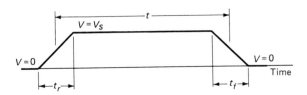

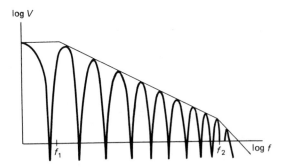

Figure 3-2   Frequency spectrum of a trapezoidal pulse

TABLE 3-1 Typical Operating Characteristics of Common Digital Integrated Circuits

Logic Family	Type	V_{cc} (V)	t_{LH} (ns)	t_{HL} (ns)	t_r (ns)*	t_f (ns)*	R_{outH} (Ω)	R_{outL} (Ω)	C_{out} (pF)	R_{in} (RΩ)	C_{in} (pF)	NML (V)	NMH (V)	I_{os} (mA)	ΔV_{out} (V)	f_2 (MHz)
40xxB	CMOS	5	125	125	80	80	600	600	10	1000	5	2.25	2.25	5	4.2	4.0
40xxB	CMOS	10	60	60	50	50	390	390	8	1000	5	4.5	4.5	25	9.0	6.4
40xxB	CMOS	15	45	45	20	20	160	160	6	1000	5	7.75	7.75	53	12.0	16
74xx	TTL	5	11	7	16	1.6	130	13	5	4	5	0.6	1.4	36	3.2	110
74ACxx	CMOS	5	5	4	2.1	2.1	31	18	10	1000	5	1.25	1.25	170	4.8	151
74ACTxx	CMOS	5	1	1	3.0	0.8	29	7	10	1000	4	0.7	2.4	140	4.3	318
74ALSxx	TTL	5	4	4	7.4	2.8	61	23	5	40	5	1.1	2.0	50	3.3	78
74ASxx	TTL	5	3	2	4.2	2.1	35	17	5	10	5	0.45	1.0	55	2.6	114
74Cxx	CMOS	5	50	50	220	220	1700	1700	10	1000	6	2.0	2.0	3	4.0	1.4
74Cxx	CMOS	10	30	30	85	75	700	600	6	10	6	5.0	3.5	16	8.0	4.0
74Fxx	TTL	5	4	3	5.6	1.5	46	12	5	10	5	1.3	1.9	90	3.3	135
74Hxx	TTL	5	6	6	7.0	1.2	58	10	5	3	5	0.6	1.5	70	3.3	155
74HCxx	CMOS	5	8	8	16	7.9	130	65	6	1000	4	0.8	1.25	45	5.0	30
74HCTxx	CMOS	5	9	9	6.1	5.5	50	45	6	1000	4	0.6	2.45	45	4.8	55
74Lxx	TTL	5	35	31	61	6.8	500	56	10	40	4	0.55	1.3	9	3.2	26
74LSxx	TTL	5	8	8	19	4.9	120	31	22	20	5	0.55	2.3	27	3.3	41
74Sxx	TTL	5	3	3	6.1	1.6	50	13	5	3	4	0.55	1.4	65	3.0	125
10xxx	ECL	−5.2	2	2	2.0	2.0	7	7	13	1	3	0.10	0.07	100	0.8	160
100xxx	ECL	−5.2	1	2	0.7	0.7	7	7	13	1	3	0.09	0.08	100	0.7	455
10G0xx	GaAs	−3.3	0.3	0.3	0.2	0.2	12	—	1	42	1	1.00	0.20	70	1.9	1600

*50 pF load capacitance.

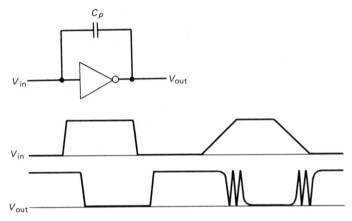

Figure 3-3 Oscillation of ICs caused by parasitic capacitance and slowly rising/falling inputs

Radiated noise is proportional to the signal voltage times the frequency (Appendix H), so f_2 is the approximate upper frequency limit for noise from trapezoidal pulses.

Digital ICs put sharp current spikes on voltage and ground lines whenever their outputs change states. Lightly loaded outputs generate $\approx I_{os}/2$ ampere spikes lasting t_r seconds, and heavily loaded outputs generate $\approx I_{os}$ ampere spikes lasting $C_L(\mid V_f - V_i \mid)/I_{os}$ seconds. Digital ICs may also oscillate at 5–50 MHz if they are poorly bypassed or the input signals change too slowly ($dV/dt < 0.1$ V/t_{LH} or 0.1 V/t_{HL} Figure 3-3). Table 3-1 summarizes all of the input/output characteristics that affect noise emissions and noise pickup in digital circuits.

RECOMMENDED READING

Design Techniques for Interference-Free Operation of Airborne Electronic Equipment. Springfield, VA: NTIS (AD 491 988), 1952.

Interference Reduction Guide for Design Engineers, Volume 1. Springfield, VA: NTIS (AD 619 666), 1964.

Zener Diode Handbook. Motorola, Inc., 1967.

DiMarzio, A. W., "Graphical Solutions to Harmonic Analysis," *1967 IEEE Electromagnetic Compatibility Symposium Record*, Washington, D.C. (July 18–20, 1967), 267–280.

Ficchi, Rocco F., ed., *Practical Design for Electromagnetic Compatibility*. New York: Hayden Book Co., Inc., 1971.

MARDIGUIAN, MICHEL, *Interference Control in Computers and Microprocessor-Based Equipment*. Gainesville, VA: Don White Consultants, Inc., 1984.

NORRIS, BRYAN, ed., *MOS and Special-Purpose Bipolar Integrated Circuits and R-F Power Transistor Circuit Design*. New York: McGraw-Hill Book Co., 1976.

THOMAS, RANDAL, and MICHAEL JENKINS, eds., *Analog Switches and Their Application*. Santa Clara, CA: Siliconix, Inc., 1980.

4

WIRING AND NOISE-COUPLING PATHS

In my fourteen years experience as a working electronics engineer *over 90 percent of the noise problems that I have encountered were caused by improper wiring!* These problems have included:

hum in home stereos,

an electronic organ picking up Citizens' Band transmissions,

oscillating amplifiers,

oscillating digital ICs,

excessive radiated noise from printed circuit boards,

digital logic reset by electrostatic discharge,

intermittent read errors with RAM memories,

multiple-clocking of counters,

shift registers losing data,

oscillating power supplies,

power supplies going out of regulation, and

stepper motors causing logic errors in a tester.

Most of these problems took days or weeks of hard work to find, and hours or days to fix. *In nearly every case, we could have avoided the problems by designing the wiring as part of the system.*

Figure 4-1 shows the lumped-impedance model for an l-meter-long wire. R_s is the wire's resistance. L_s is parasitic inductance, and C_p is parasitic capacitance (Appendices D and E), which together cause an end-to-end propagation delay of $(L_s C_p)^{1/2}$ seconds. V_s is noise induced on the wire by electromagnetic fields, thermoelectric effects, and galvanic effects.

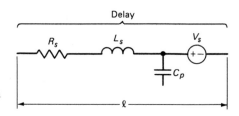

Figure 4-1 Lumped-impedance model of a wire

Figure 4-2 illustrates *common-impedance coupling* caused by wiring resistance. Figure 4-2(a) shows the circuit schematic—U1 and U2 form a quiet circuit, and V_a and R_L form a noisy circuit. Most electronics books define ground as "a zero-reference point," so we expect U2 to see V_q volts on its noninverting input, and we feel safe connecting the ground pins to any available ground. Figure 4-2(b) shows one likely result—U1 and U2 share part of the ground net with V_a and R_L. When we test this circuit, we find $(V_a R_s)/(R_L + R_s)$ volts of *common-impedance-coupled noise* on U2's input. We can reduce this noise by decreasing R_s, or we can eliminate the noise entirely by splitting the ground net into two separate branches (R_{s1} and R_{s2} in Figure 4-2(c)).

Lead inductance strongly affects the high-frequency behavior of low-value resistors. The lead inductance (L_s henries) and the resistor's parasitic

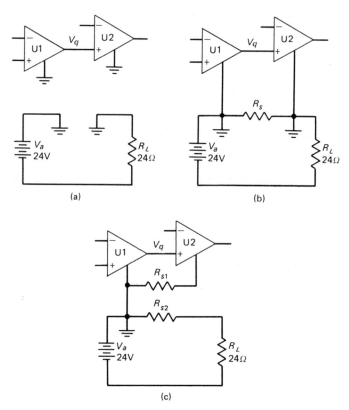

Figure 4-2   Common-impedance coupling through wiring resistance: (a) schematic, (b) wiring scheme 1, (c) wiring scheme 2

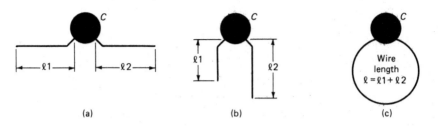

Figure 4-3   Self-resonance of a capacitor: (a) straight leads, (b) parallel leads, (c) curved leads

capacitance (C_p farads) resonate at $f_c = 1/[2\pi(L_sC_p)^{1/2}]$ hertz, producing large impedance shifts over a narrow frequency band (see Figure 2-3).

Lead inductance has its most serious effect on capacitors. As shown back in Chapter 2, the self-resonant frequency of a capacitor, $f_c = 1/[2\pi(L_sC)^{1/2}]$ hertz, is the upper frequency limit for capacitive behavior. L_s is the sum of the capacitor's internal inductance, lead inductance, and circuit inductance. For example, high-frequency bypass capacitors are capacitive up to $f_c/3$ hertz, near short circuits from $f_c/3$ hertz to $3f_c$ hertz, and inductive above $3f_c$ hertz. Straight or parallel lead wires (Figure 4-3(a), (b)) act like a single piece of wire $l = l1 + l2$ meters long. Curved lead wires (Fig. 4-3(c)) act like a circle of wire with circumference $l = l1 + l2$ meters (see Appendix D).

Table 4-1 shows calculated self-resonant frequencies for 100 pF to 10 µF capacitors with 3.2-mm- to 25.4-mm-long straight leads. Table 4-2 shows calculated and measured self-resonant frequencies for twelve capacitors with their leads bent into circles. As this table shows, we can estimate the self-resonant frequency of a capacitor with ± 10% accuracy (calc. f_c vs. HP4191 f_c) just from the physical dimensions of the leads. Using an inexpensive ($\approx$$100) grid-dip meter, and following the pro-

TABLE 4-1   Self-Resonant Frequency of Capacitors with Straight Leads

Capacitance	$l = 3.2 + 3.2$ mm $L_s = 3.76$ nH	$l = 12.7 + 12.7$ mm $L_s = 22.0$ nH	$l = 25.4 + 25.4$ mm $L_s = 50.9$ nH
100 pF	260 MHz	107 MHz	70.5 MHz
1000 pF	82.1 MHz	33.9 MHz	22.3 MHz
0.01 µF	26.0 MHz	10.7 MHz	7.05 MHz
0.1 µF	8.21 MHz	3.39 MHz	2.23 MHz
1 µF	2.60 MHz	1.07 MHz	0.71 MHz
10 µF	0.82 MHz	0.34 MHz	0.22 MHz

TABLE 4-2   **Self-Resonant Frequency of Capacitors with Curved Leads**

Capacitor Type	C		$l1 + l2$	D	Calc. L_s	Calc. f_c	Grid-dip f_c	HP4191 f_c (ref)
Disc ceramic	1027	pF	40 mm	0.64 mm	26.6 nH	30.4 MHz	31.3 MHz	32 MHz
Mylar	1851	pF	40 mm	0.64 mm	26.6 nH	22.7 MHz	23.3 MHz	23 MHz
Tubular ceramic	8.68	nF	20 mm	0.51 mm	11.4 nH	16.0 MHz	16.5 MHz	16 MHz
Tubular ceramic	10.6	nF	40 mm	0.51 mm	28.4 nH	9.2 MHz	9.5 MHz	9.6 MHz
Tubular ceramic	9.25	nF	80 mm	0.51 mm	67.9 nH	6.3 MHz	6.7 MHz	7.0 MHz
Tubular ceramic	10.7	nF	20 mm	0.51 mm	11.4 nH	14.4 MHz	14.1 MHz	15 MHz
Tubular ceramic	10.8	nF	40 mm	0.51 mm	28.4 nH	9.1 MHz	9.2 MHz	9.4 MHz
Disc ceramic	11.5	nF	40 mm	0.64 mm	26.6 nH	9.1 MHz	8.3 MHz	8.6 MHz
Disc ceramic	14.2	nF	80 mm	0.64 mm	64.4 nH	5.3 MHz	5.5 MHz	5.6 MHz
Tantalum elec.	99.8	nF	40 mm	0.51 mm	28.4 nH	3.0 MHz	≈ 3 MHz	11 MHz*
Tubular ceramic	106	nF	40 mm	0.51 mm	28.4 nH	2.9 MHz	3.0 MHz	3.0 MHz
Mylar	216	nF	40 mm	0.64 mm	26.6 nH	2.1 MHz	2.0 MHz	1.9 MHz

*Extremely lossy, 40-degree phase angle below resonance.

cedure in Chapter 2, we can measure the self-resonant frequency with roughly ±5% accuracy (grid-dip f_c vs. HP4191 f_c).

Inductive coupling mainly affects low-impedance circuits (loop impedance < 376.7 Ω, the intrinsic impedance of air). Wires, open-core inductors, and transformers pick up or generate magnetic fields, producing "hum" in audio systems and crosstalk/electromagnetic interference (EMI) in digital systems. Inductive coupling can be a serious problem when sensitive circuits are close to high-current circuits, a common occurrence with interplane boards (see Appendix G).

Capacitive coupling mainly affects high-impedance circuits (loop impedance > 376.7 Ω), when wires and other ungrounded pieces of metal pick up/generate electric fields. Capacitive coupling is a common problem in noninterplane boards and analog circuits (see Appendix G). Wiring capacitance can seriously affect tuned circuits. Figure 4-4 shows

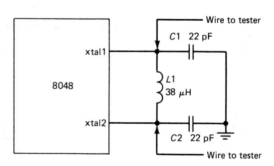

Figure 4-4   Wiring capacitance upsetting a tuned circuit

a clock circuit that was designed to oscillate at 5.8 MHz. This circuit oscillated at 2 MHz when connected to a tester because of the tester's wiring capacitance. (I wound up removing the wires completely. In my final test, I assumed that $C1$, $C2$, and $L1$ were okay if the 8048 finished the test in the allotted time.) Wiring capacitance can also make digital ICs oscillate, as illustrated back in Figure 3-3.

Signal-propagation delays on wires can cause glitches (unwanted voltage spikes) and other hard-to-find problems. Figure 4-5 shows a 74LS93 counter with long wires going to a tester. About 85% of the 74LS93s tested would count correctly, about 5% would count correctly some of the time, and about 10% always counted too fast. A fast oscilloscope revealed $\approx 1V$ glitches on QB and QC (see Appendix F), which would occasionally toggle the next flipflop in the string and cause a miscount. Disconnecting the wires from QB and QC eliminated the glitches but left us without a test. We finally solved the problem by installing special buffer boards in the test fixture, so the 74LS93s would drive 74LS09 gates through $\approx 0.15m$-long wires, and the 74LS09s would drive the 2m-long tester wires. The tester also had problems with 74LS96 shift registers (Figure 4-6): the long tester wires caused $\approx 1V$ overshoot on falling edges, which would occasionally reset the following flipflop. We solved this problem the same way, adding buffer boards so that the 74LS96s drove only $\approx 0.15m$-long wires.

Propagation delays in the wiring can also cause power supplies to oscillate. We have seen this problem numerous times when trying to test printed circuit boards that lack bulk bypass capacitors. In another case, a linear power supply's $+5$-V output would randomly jump between $+5.1$ V and $+5.6$ V because of oscillations on its DC input. We solved these problems by adding bulk bypass capacitors to the test fixtures in order to keep current spikes from reaching the tester's power supply.

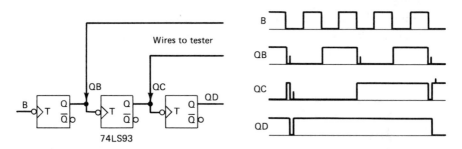

Figure 4-5   Counter errors caused by long wires

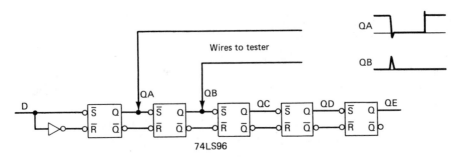

Figure 4-6   Shift-register errors caused by long wires

A wire, just by its size and shape, also forms an antenna that can pick up or emit electromagnetic noise (see Appendix H). Wires longer than $\lambda/8$, where λ is the signal's wavelength ($\lambda \approx 3 \times 10^8/f_2$ meters, f_2 in hertz), form efficient antennas. For systems using 74LS and 74S logic, wires over 0.3 meter long can be major noise sources. In addition, points l-meters apart on a conductor see an impedance proportional to $1 + |\,\text{etan}\,(2\pi l/\lambda)\,|$ because of standing waves. In effect, points separated by 0.25λ, 0.75λ, 1.25λ, etc. are completely isolated from one another. This effect limits the size of ground systems: Most circuits can stand $\leq 0.10\lambda$ separation between grounds, and insensitive circuits can have grounds 0.15λ apart, but sensitive circuits must have all their grounds within 0.05λ of one another.

Most system noise comes from common-impedance coupling, inductive coupling, capacitive coupling, and antenna effects. Vibrating wires can also induce noise through triboelectric effects (different materials rubbing together) and magnetic induction. Galvanic effects may be important if two metals are separated by a wet porous insulator, forming a small battery. Some circuits are even sensitive to thermoelectric effects (the Seebeck effect) caused by temperature differences across metal-to-metal junctions.

Figure 4-7 shows an integrated circuit on a printed circuit board. The copper-solder-kovar bond can generate 37 μV/°C thermoelectric voltage. This voltage can be reduced to ≈ 0.5 μV/°C by using ICs with copper

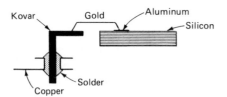

Figure 4-7   Contacts between different metals on a printed circuit board

or Alloy 42 leads. To reduce the thermoelectric voltage even more, we can use 30% tin/70% cadmium solder instead of regular tin-lead solder (0.7 μV/°C versus 3 μV/°C for the solder-to-copper bond). We can also thermally match differential circuits by keeping the signal path and the signal-return path close together and routing them through matched components.

RECOMMENDED READING

Bell Telephone Laboratories, *Physical Design of Electronic Systems,* Vol. I. Englewood Cliffs, NJ: Prentice-Hall, 1970.

OTT, HENRY W., *Noise Reduction Techniques in Electronic Systems.* New York: John Wiley & Sons, 1976.

PASCOE, G., "The Thermo-EMFs of Tin-Lead Alloys," *Journal of Physics E: Scientific Instruments,* **9** (1976), 1121–1122.

5

DESIGNING ANALOG CIRCUITS

Most analog circuits employ low-level signals and thus tend to be noise victims. Analog circuits should be designed for linear operation with the minimum required gain and bandwidth. Noise pickup can be reduced by using differential signals, keeping output impedances under 1 kΩ, and keeping load impedances over 300 Ω. High-gain amplifiers tend to oscillate at frequencies between 10 kHz and 5 MHz, so feedback loops should be designed to prevent these oscillations under worst-case conditions. If high-level noise does enter an analog circuit, it can change the biasing and thus desensitize or overload the amplifiers.

Analog circuits need effective bypassing and decoupling to limit noise pickup through the power lines. Figure 5-1 shows the recommended bypassing for operational amplifiers (opamps). V+ and V− should each have one 1-to-10 μF tantalum electrolytic bypass capacitor per five opamps, spread throughout the circuit. Each opamp should have ceramic bypass capacitors connecting its V+ and V− pins to the output signal's return line. These bypass capacitors should be 0.1 μF or at least 100 times the load capacitance, whichever is greater. Insufficient bypassing often shows up as oscillations or motorboating ("putt-putt" sounds). (*Note:* If you parallel large and small bypass capacitors on a net, you may want to put a ≈1Ω resistor in series with the large capacitor to reduce high-frequency ringing.)

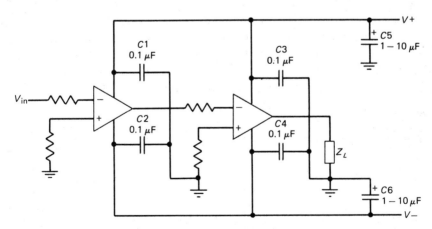

Figure 5-1 Recommended bypassing for opamps

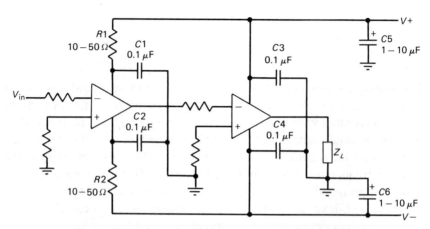

Figure 5-2   Recommended decoupling for multistage amplifiers

Figure 5-2 shows the recommended decoupling for multistage amplifiers. $R1$ and $R2$ help keep power-line noise from coupling into the first stage. The power input is close to the output stage in order to decrease power-line noise into the input stage and thus reduce the chance of oscillations. The ideal layout for a multistage amplifier is a straight line, keeping the input stage and the output stage as far apart as possible.

If an opamp directly drives a reactive load (L_L and C_L in Figure 5-3), it is almost guaranteed to oscillate. A small series damping resistor (Fig. 5-3(a), $R_L \geqslant 2(L_L/C_L)^{1/2}$ ohms) or a ferrite bead on the opamp's output lead (Fig. 5-3(b)) can tame these wild oscillations.

Opamps may also oscillate if they have to drive capacitive loads. Figure 5-4 shows two ways to keep inverting amplifiers from oscillating. $R1$ and $R2$ set the amplifiers' gain. $R3 \approx (R1 \cdot R2)/(R1 + R2)$ ohms is optional, but it helps balance the opamps' input bias currents and if kept close to $R1$ will help cancel any thermoelectric voltage induced in the input circuitry. In Figure 5-4(a), adding $C1 \geqslant 15(R1/R2)$ pF makes the amplifier stable under almost all loading conditions. In Figure 5-4(b) we

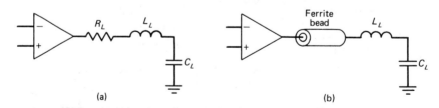

Figure 5-3   Driving reactive loads with opamps

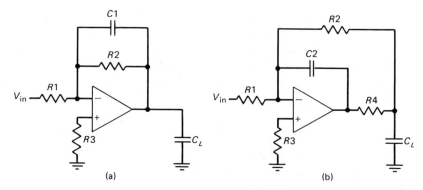

Figure 5-4   Driving capacitive loads with inverting amplifiers

add resistor $R4$, which is much greater than the opamp's output resistance, and $C2 \geq C_L(R4/R2)$ to make the amplifier stable under all loading conditions.

Figure 5-5 shows four ways to keep noninverting amplifiers from oscillating. In Figure 5-5(a), $R5$ and $C3$ slow down the input signal so that the opamp can charge C_L without saturating. In Figure 5-5(b), $R6$

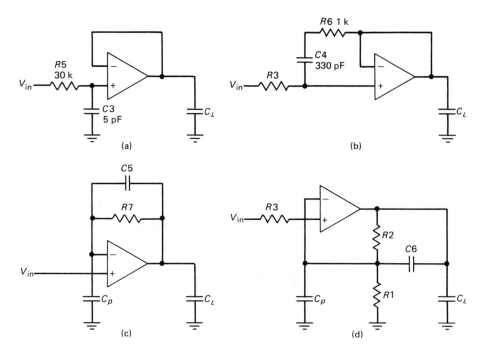

Figure 5-5   Driving capacitive loads with noninverting amplifiers

and $C4$ reduce the opamp's high-frequency gain without affecting its DC gain. In Figure 5-5(c), $R7$ and $C5 \gg C_p$ (where C_p is parasitic capacitance on the node) make the opamp into an integrator, which is unconditionally stable. In Figure 5-5(d), adding $C6 \approx C_p$ $(R1/R2)$ creates a capacitive divider in parallel with the resistive divider ($R1$ and $R2$) in the feedback loop. An excellent way to check an amplifier for stability is to hook a pulse generator to the input and adjust it for a $\approx$200-mV step on the amplifier's output (with the normal loads connected). If the output has less than 40 percent overshoot, the circuit is stable.

The inputs of some analog circuits can be switched between various signal sources (i.e., phonograph, radio, and cassette deck in a home stereo). AC-coupled inputs should have pull-down resistors ($R1$ and $R2$ in Figure 5-6(a)) to keep the input capacitors discharged and thus prevent clicks, pops, and other transients. FET switches can couple the drive signal into the analog input through the parasitic gate-to-drain capacitance. This noise can be reduced by reducing the swing of the gate signal or by adding an R-C filter to the gate (Figure 5-6(b)) to stretch out the gate signal. If the signals must be switched very rapidly, a differential amplifier

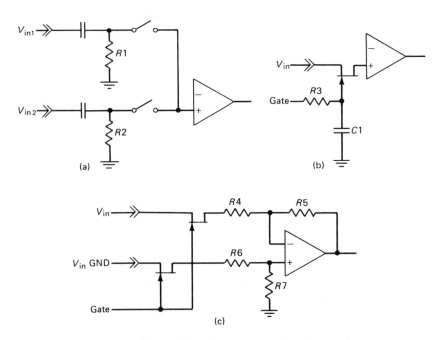

Figure 5-6   Switching analog signals

with matched FET switches and matched inputs ($R4 = R6$, $R5 = R7$ in Figure 5-6(c)) will cancel out the charge transfers.

In 1978 I was designing a functional tester for a new ink-jet printer. This tester had to monitor about 30 critical analog signals, ranging from -12 V to $+300$ V, with analog-input boards rated for ± 5 V. To protect the expensive analog boards I decided to buffer all the analog signals with inexpensive opamps. Figure 5-7 shows my original designs for a noninverting amplifier (for signals under ± 4 V) and for an inverting amplifier (for signals over ± 4 V). From the preliminary mechanical layout, I expected the output cables to be about two meters long.

As part of my design, I breadboarded the circuits and tested them with capacitive loads from 10 pF to 0.1 μF, both with and without seven meters of twisted-pair cable between the opamp and the load. My test procedure was to toggle the input between $+5$ V and ground while observing the opamp output on an oscilloscope. Experimenting with no bypass capacitors, 0.1-μF capacitors, and 0.68-μF capacitors, I got ringing or oscillations for certain ranges of output capacitance, but the 0.68 μF bypass capacitors seemed to work best. Further experiments showed that adding a small feedback capacitor would stabilize the inverting amplifier, and adding a low-pass input filter would stabilize the noninverting buffer. So I designed the tester using the circuits in Figure 5-8.

Some six months later, while debugging the tester, I encountered one noise problem—sometimes one noninverting buffer would oscillate for a few milliseconds when the input toggled between $+5$ V and -5 V. Changing $R5$ and $C3$ didn't help, so I tried adding a resistor between the opamp output and the twisted-pair cable. The lowest value that worked consistently was 27 Ω, so I installed a 47-Ω resistor in the circuit to provide some safety margin. Three copies of this tester have been in daily

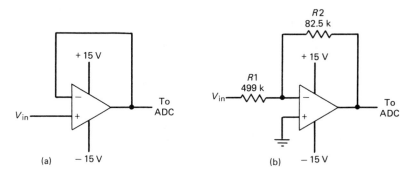

Figure 5-7 Conditionally stable amplifiers

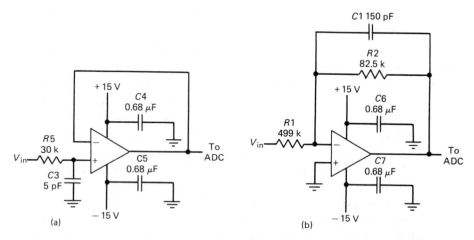

Figure 5-8 Stable amplifiers

use for seven years now, testing thousands of printers with no noise problems whatsoever.

RECOMMENDED READING

BROKAW, PAUL, "An IC Amplifier User's Guide to Decoupling & Grounding or . . . Making Things Go Right," *Electronic Products Magazine*, **20**:7 (December 1977), 44–53.

FICCHI, ROCCO F., ed., *Practical Design for Electromagnetic Compatibility*. New York: Hayden Book Co., Inc., 1971.

GILES, MARTIN, ed., *Audio/Radio Handbook*. Santa Clara, CA: National Semiconductor Corp., 1980.

JONES, DON, "Operational Amplifier Stability: Input Capacitance Considerations." Melbourne, FL: Harris Semiconductor Application Note 515.

STOUT, DAVID F., and MILTON KAUFMAN, eds., *Handbook of Operational Amplifier Circuit Design*. New York: McGraw-Hill Book Co., 1976.

6

DESIGNING DIGITAL CIRCUITS

Digital ICs tend to be both producers and victims of electrical noise. Table 6-1 summarizes the noise characteristics of common digital integrated circuits (ICs). Whenever a digital signal toggles, it generates significant noise from DC up to f_2 megahertz. During the transition the output voltage changes at dV/dt volts/nanosecond, producing crosstalk, and a transient pulse of at least V_{cc}/GND-spike picocoulombs of charge flows from V_{cc}, through the driver, and into ground. To prevent glitches, unterminated transmission lines must be kept less than $l_c = 0.5t_r/t_u$ meters long. Similarly, the noise on input signals must be less than the noise margin to prevent system malfunctions.

To reduce noise emissions from digital logic, keep signals slow (long rise and fall times) with small voltage swings, limit the number of signals that switch at one time, and use good grounding and bypassing techniques. To reduce noise pickup, use slow synchronous ICs with Schmitt-trigger inputs. If the system has long cables, use differential

TABLE 6-1   Noise Characteristics of Common Digital ICs

Logic Family	V_{cc} (V)	f_2 (MHz)	dV/dt (V/ns)	V_{cc}/GND Spike (pC)	l_c (m)	Noise Margin (V)
40xxB	5	4.0	0.05	170	6.06	2.25
40xxB	10	6.4	0.18	230	3.79	4.50
40xxB	15	16	0.60	120	1.52	7.75
74xx	5	110	2.00	26	0.12	0.60
74ACxx	5	151	2.29	5.0	0.16	1.25
74ACTxx	5	318	5.38	6.5	0.06	0.70
74ALSxx	5	78	1.18	12	0.21	1.10
74ASxx	5	114	1.24	5.5	0.16	0.45
74Cxx	5	1.4	0.02	440	16.67	2.00
74Cxx	10	4.0	0.11	340	5.68	3.50
74Fxx	5	135	2.20	9.2	0.11	1.30
74Hxx	5	155	2.75	12	0.09	0.60
74HCxx	5	30	0.63	40	0.60	0.80
74HCTxx	5	55	0.87	15	0.42	0.60
74Lxx	5	26	0.47	98	0.52	0.55
74LSxx	5	41	0.67	31	0.37	0.55
74Sxx	5	125	1.88	9.2	0.12	0.55
10xxx	−5.2	160	0.40	0.80	0.15	0.07
100xxx	−5.2	455	1.00	0.25	0.05	0.08
10G0xx	−3.3	1600	9.50	0.19	0.02	0.20

drivers and differential receivers connected by balanced transmission lines to reduce noise emissions and noise pickup.

The 74Cxx and 40xxB logic families are excellent for low-noise systems. They are quiet, have wide noise margins, and can safely drive unterminated lines up to 3.7 meters long (assuming $t_u \leqslant 6.6$ ns/m). Second best are the 74HCxx, 74HCTxx, 74Lxx, and 74LSxx logic families, able to drive up to 0.37-meter-long unterminated transmission lines. The 74xx, 74ACxx, 74ALS, 74ASxx, 74Fxx, 74Hxx, 74Sxx, and 10xxx logic families require unterminated transmission lines to be less than 0.09 meter long. The 74ACTxx, 100xxx, and 10G0xx logic families are extremely noisy and require almost all signal lines to be properly terminated.

Bypass capacitors supply the transient current needed by digital ICs, reduce voltage drops in the V_{cc} and ground nets, and help filter out power/ground noise. Figure 6-1(a) shows the traditional brute-force method for bypassing digital circuits. A large electrolytic bypass capacitor (10–100 μF, $\geqslant 1$ μF per IC) is placed near the power input. Each edge-triggered IC gets a 0.1-μF ceramic bypass capacitor, and high-speed ICs also get small (100 to 1000 pF) ceramic bypass capacitors. Each IC that drives or receives off-board signals gets a 0.1-μF ceramic bypass capacitor between V_{cc} and the signal return (Figure 6-2). The rest of the ICs get a sprinkling of 0.01-μF to 0.1-μF ceramic bypass capacitors, with at least one capacitor per five ICs. This bypassing scheme has two problems: the bypass capacitors are expensive, and they do a poor job of filtering out noise above 10 MHz because of their low self-resonant frequencies (see Table 4-1).

Figure 6-1(b) shows the bypassing scheme recommended by R. Kenneth Keenan in all three of his books. $C6$, $C7$, and the ferrite bead form a pi-filter to attenuate high-frequency noise leaving the board. Each IC that drives or receives off-board signals gets a 0.1-μF ceramic bypass capacitor between V_{cc} and the signal return (Figure 6-2). All other ICs get one ceramic bypass capacitor each, with capacitance $C \geqslant (4.5)(\#$ of outputs$)(C_L)$ farads. A tantalum electrolytic (or metallized polycarbonate) capacitor is placed near the power input to filter low-frequency noise. This capacitor should have at least ten times as much capacitance as all the other capacitors on the net combined.

This arrangement provides good bypassing up to 100 MHz+ using inexpensive components. The ceramic capacitors should have equivalent series inductance (*ESL*) < 20 nH and equivalent series resistance (*ESR*) < 0.5 Ω. Including lead inductance, the tantalum electrolytic (or polycarbonate) capacitors should have *ESL* < 30 nH and *ESR* < 1 Ω. The

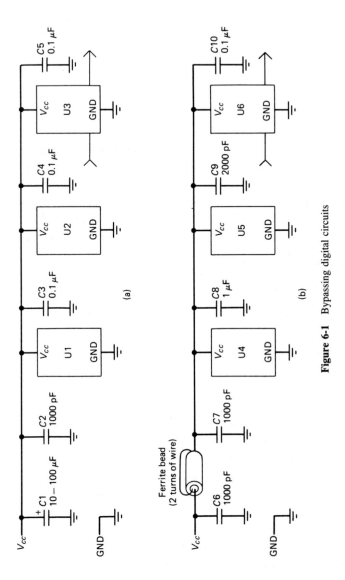

Figure 6-1 Bypassing digital circuits

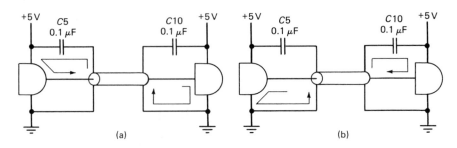

Figure 6-2 Bypassing off-board drivers and receivers: (a) low-to-high transition, (b) high-to-low transition

ferrite beads should have $< 10\ \Omega$ impedance at the circuit's clock frequency and $> 50\ \Omega$ impedance at five times the clock frequency (at the circuit's operating current). For best results the wire should pass through the ferrite bead two times. If a single bead's impedance is too low, use several beads in series or choose a longer, fatter bead. If the circuits can stand the voltage drop, $\approx 51\text{-}\Omega$ carbon-composition or metal-film resistors may be used instead of ferrite beads.

Clock signals, and their harmonics, dominate radiated emissions between 30 MHz and 1 GHz. The even harmonics can be greatly reduced by using clocks with $\approx 50\%$ duty cycles (''on'' times approximately equal to the ''off'' times). Try to limit the number of ICs served by each clock. If clocks must go to several boards, use Schmitt-trigger-input gates as buffers, and limit the voltage swing and the edge rate (dV/dt) of the main clock signals. If clocks must be controlled by off-board switches, do not route the clocks through the switches. Instead let the switches operate control lines to on-board gates that in turn enable/disable the clocks. Staggering the clock signals, and using multiple oscillators, will also help reduce clock noise.

Careful system timing can also reduce noise problems. To reduce transient currents on V_{cc} and ground, use staggered clocks to control small groups of ICs. Use synchronous ICs and input data strobes to reduce the percentage of the time that a system is sensitive to noise.

Each input signal to a board should go to only one IC (preferably one with Schmitt-trigger inputs) to prevent timing problems. Schmitt-trigger inputs increase the noise immunity and easily handle slow signals. If an input signal must go to a regular gate, keep the rise and fall times short to prevent oscillations ($dV/dt > 0.1V/t_{HL}$ and $dV/dt > 0.1V/t_{LH}$).

Signals that go off-board require special care. The outputs of flip-flops, counters, and shift registers should be buffered through gates or

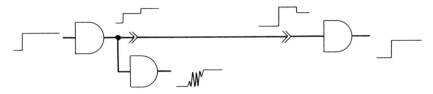

Figure 6-3   Glitches caused by reflections on long wires

line drivers to prevent glitch problems (see Figures 4-5 and 4-6). Stringing an output wire through a ferrite bead will allow a longer-than-normal unterminated line to be driven safely. *Signals going off-board should not drive on-board inputs!* Violating this rule can easily create severe noise problems at the driver (Figure 6-3).

Long transmission lines must be terminated in their characteristic impedance to prevent reflections and glitches (see Chapter 13 and Appendix F). A series resistor at the driver will work if the receivers are all at the far end of the line (Figure 6-4(a)). The resistor should leave the line slightly underdamped, with just a little overshoot when the signal toggles. A resistive divider at the far end of the line will allow receivers to be placed anywhere along the line (Figure 6-4(b)). To achieve high noise rejection on long cable runs, use differential drivers and receivers with balanced lines. For unidirectional lines, the far ends must be terminated (Figure 6-5(a)). For bidirectional lines, both ends must be terminated (Figure 6-5(b)).When properly terminated, balanced lines can achieve 70 dB (3000:1) noise rejection from DC to 100 kHz.

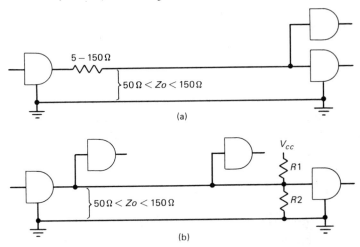

Figure 6-4   Terminating long transmission lines to reduce reflections

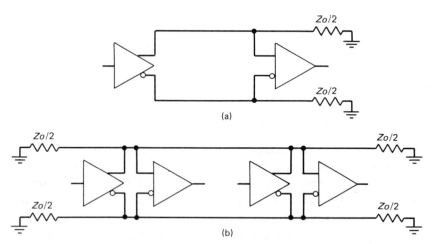

Figure 6-5   Driving balanced transmission lines: (a) unidirectional line, (b) bidirectional line

Cables, PCB lands, and wires should be terminated whenever the signal propagation time exceeds half the signal rise or fall time. Flat cables, twisted-pair cables, and coaxial cables should be terminated in ≈ 100 Ω ($75\Omega \leqslant Zo \leqslant 120$ Ω). PCB lands should be terminated in their characteristic impedance (20 $\Omega \leqslant Zo \leqslant 200$ Ω). A resistive divider (Figure 6-6(a)) can terminate and bias a line without requiring additional power supplies. For example, standard terminators for TTL logic are $R1 = 330$ Ω and $R2 = 220$ Ω, giving 132 Ω into $+2$ V ($V_{cc} = +5$ V); $R1 = 150$ Ω and $R2 = 470$ Ω, giving 114 Ω into $+3.8$ V; and $R1 = 120$ Ω and $R2 = 470$ Ω, giving 96 Ω into $+4.0$ V. Most CMOS ICs cannot drive these resistive dividers. Instead they may use the resistive/capacitive divider shown in Figure 6-6(b), with $R3 \approx 1$ kΩ, $R4 \approx 330$ Ω, and $C1 \approx 1000$ pF.

Critical applications may require a tighter impedance match between the transmission lines and their terminating networks. Robert Barnes (my father) devised a method to find the best terminating network even when the impedance is unknown, the impedance varies with frequency, or parasitic capacitance/inductance affects the impedance at the termination. His method is as follows:

1. Build a prototype.
2. Choose the no-load voltage, $V0$, for the network. For maximum speed, choose $V0 \approx V_{cc}/2$. Choose V_{in} (high) $< V0 < V_{cc}$ to guar-

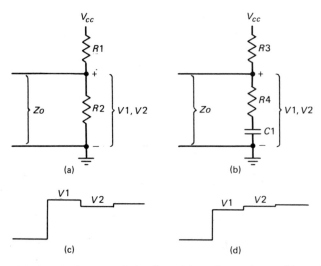

Figure 6-6 Terminating transmission lines: (a) regular terminator, (b) terminator for CMOS, (c) $R_t < Zo$, (d) $R_t > Zo$

antee a "1" on an idle line. Choose $0 < V0 < V_{in}$ (low) to guarantee a "0" on an idle line.

3. Calculate or assume a value for Zo.
4. Select 1% metal-film resistors for $R1$ and $R2$, with $R1 \approx ZoV_{cc}/V0$ ohms and $R2 \approx ZoR1/(R1 - Zo)$ ohms.
5. Mount $R1$ and $R2$ in the prototype.
6. Using the prototype's driver, send a pulse down the line and observe the waveform at the driver with a fast oscilloscope. Measure $V1$ and $V2$, the initial output voltage and the first reflected voltage (Figure 6-6(c) and (d)).
7. The transmission-line impedance *at the terminator* is

$$Zo = \left(\frac{R1 \cdot R2}{R1 + R2} \right) \left(\frac{2V1 - V2}{V2} \right) \text{ ohms.}$$

8. Using this value for Zo, repeat steps 4 through 6. This time you should not see a reflection, indicating a perfect impedance match at the termination.

All unused IC inputs should be pulled up or pulled down. Some systems can turn off subsystems that aren't in use. If these subsystems

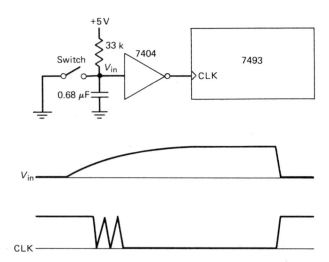

Figure 6-7   Counting errors caused by slow-rise-time signals

use TTL logic, the inputs should be active-high with pull-down resistors, and the outputs should be designed to stay high during power-up and power-down sequences.

In 1977 one of my co-workers was having trouble with a small engineering tester for a new ink-jet printer. The circuit in Figure 6-7 was supposed to record how many times the printer carriage bashed into the left frame. The 0.68-μF capacitor had been added to the circuit to eliminate switch bounce, but the counter still jumped by two or three counts every time the switch opened. Poking around with an oscilloscope, we saw the 7404's output oscillating, because of the long rise time of V_{in}. We cured the problem by replacing the 7404 with a 7414.

RECOMMENDED READING

KEENAN, R. KENNETH, *Decoupling and Layout of Digital Printed Circuits*. Pinellas Park, FL: TKC, 1985.

KEENAN, R. KENNETH, *Digital Design for Interference Specifications*. Pinellas Park, FL: TKC, 1983.

KEENAN, R. KENNETH, *FCC/VDE Noise Specifications: Application of Ferrite Beads to Decoupling Printed-Circuit Boards from Backplanes*. Pinellas Park, FL: TKC, 1984.

OTT, HENRY W., *Noise Reduction Techniques in Electronic Systems*. New York: John Wiley & Sons, 1976.

7

DESIGNING
INTERFACE
CIRCUITS

The interface between an electronic system and the "real world" tends to be a major noise source. Input/output devices are often separate units, connected to the main system by long cables. Many input/output devices include solenoids, switches, relays, and motors, which operate at high power levels and generate large voltage and current spikes. To reduce these noise problems, design interface circuits for slow, voltage- and current-limited operation, use arc- and spike-suppression circuits, and keep the high-power circuits as compact as possible.

Nonlinear junctions can rectify high-frequency noise and inject it into low-frequency circuits, an effect called "audio rectification." Business machines may see 10-V/m electric fields, and military equipment 100-V/m fields, so long unshielded cables can easily pick up over 10 V of common-mode noise. At this level, not only the PN-junctions in transistors and ICs, but even cold-solder joints and corroded connections can act as rectifiers. Audio rectification can be reduced by modifying the wiring to reduce noise pickup or adding low-pass filters to protect sensitive junctions (Figure 7-1).

Output circuits must be designed to supply the high inrush currents drawn by their loads. The turn-on current for an incandescent lamp, for example, may be 10 to 15 times its operating current. At turn-on a transformer can draw 100 times its operating current, a motor 25 times, and a relay 15 times its operating current. These heavy inrush currents generate a lot of noise and can weld contacts or destroy semiconductors.

SCRs and triacs produce lots of noise because they turn on quickly

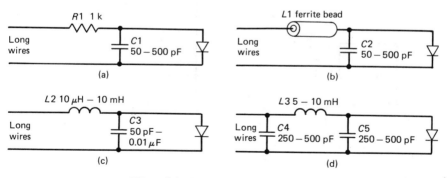

Figure 7-1   Reducing audio rectification

and drive heavy loads. This noise can be reduced by turning them on only at zero-crossings of the supply voltage. Sharp voltage spikes at the anode of an SCR or triac can couple into the gate through parasitic capacitance and accidentally turn on the device. RC-snubbers (Figure 7-2, (a) through (d)) absorb these spikes to prevent accidental turn-on. As a rule of thumb, the resistor in the snubber should equal the minimum load resistance. For small- to medium-current SCRs or triacs driving loads with inductance L_L (henries), $10 \ \Omega \leqslant R \leqslant 200 \ \Omega$ and $L_L/R^2 \leqslant C \leqslant 4L_L/R^2$ farads. If the load resistance is less than a couple of ohms, $R4$ and CR4 can be omitted from the snubber in Figure 7-2(d).

Bipolar transistors and field-effect transistors (FETs) may oscillate because of parasitic capacitance to their bases and gates. This is especially common when high-frequency transistors (unity-gain frequency $f_t \geqslant 100$ MHz) are operated below $0.2f_t$ hertz. We can prevent high-frequency oscillations in bipolar transistors by placing 10 to 100pF base-emitter capacitors ($C1$ and $C2$ in Figure 7-3) close to the bipolar transistors. For FETs, 100Ω to $2k\Omega$ resistors in series with the gates ($R1$ and $R2$ in Figure 7-3) have the same results. Another effective technique, which doesn't require artwork changes, is to put ferrite beads on the base/gate leads of

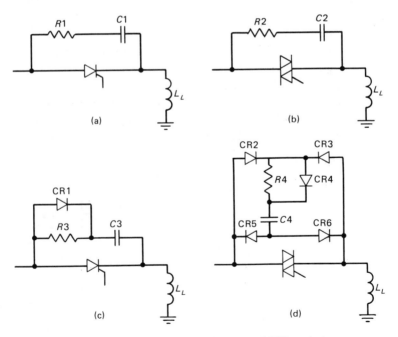

Figure 7-2 Preventing accidental turn-on of SCRs and triacs

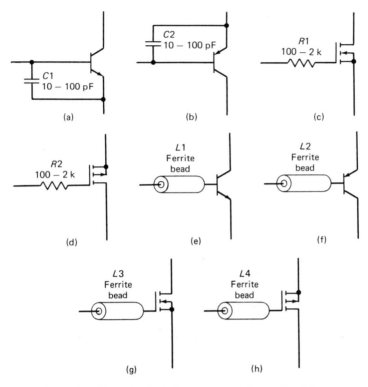

Figure 7-3   Preventing high-frequency oscillations in transistors

the transistors ($L1$ to $L4$ in Figure 7-3) to reduce the high-frequency gain without affecting the low-frequency characteristics. The fast turn-on/turn-off of transistors and FETs may cause excessive radiated noise. A ≈ 0.047 µF capacitor from the collector to the emitter (drain to source) will slow these edges and reduce the radiated noise. A better solution is to put the drivers very close to the loads in order to reduce the loop areas and lengths of wires carrying heavy transient currents.

Arc-suppression circuits reduce the noise produced by the opening and closing of switches and relay contacts, especially when driving inductive loads. Glow discharge occurs when the voltage across contacts exceeds 300 volts. Arcing occurs when (1) the voltage across the contacts changes faster than 1 V/µs; (2) the voltage exceeds the V_{arc} rating of the negative contact; and (3) the load current exceeds the I_{arc} rating of either contact. Table 7-1 lists V_{arc} and I_{arc} ratings for common contact materials. Contacts with a high V_{arc} rating also tend to produce less electromagnetic interference (EMI) than contacts with a low V_{arc} rating, because of the

TABLE 7-1   Minimum Arcing Conditions for Common Contact Materials

Contact Material	V_{arc} (V)	I_{arc} (A)
Carbon	15.5–20	0.01–0.03
Copper	8.5–14	0.36–0.60
Gold	9–16	0.38–0.42
Iron	8–13	0.35–0.73
Molybdenum	17	0.75
Nickel	8–14	0.20–0.50
Palladium	15–16	0.80
Platinum	13.5–17.5	0.67–1.00
Rhodium	14	0.35
Silver	8–13	0.40–0.90
Tungsten	10–16.5	0.90–1.27

reduced arcing. (*Note:* Contacts previously damaged by arcing may have a minimum arcing current of one-tenth the listed value.)

Figure 7-4 shows some common arc-suppression networks for switches and relay contacts. Resistive loads drawing less than I_{arc} amperes do not need arc suppressors. Figures 7-4(a), (b), (c) and (d) show arc suppressors for inductive loads carrying less than I_{arc} amperes. Figures 7-4(e) and (f) show arc suppressors for loads carrying more than I_{arc} amperes. Assuming a peak supply voltage of V_s volts, peak load current of I amperes, a load inductance of L_L henries, and a load resistance of R_L ohms, the recommended values are

C1, C2, C3, C4 and C5 $\geqslant 10^{-6}\,I$ farads and $\geqslant (I/300)^2 L_L$ farads, with a working voltage of 10 V_s volts;

C6 and C7 $\approx L_L/R_L^2$ farads;

CR1, CR2, and CR3 rated for V_s volts peak inverse voltage and I amperes continuous current;

L1 $\approx$ 10 µH;

V/I_{arc} ohms $\leqslant R1 \leqslant R_L$ ohms;

R2 and R3 $\leqslant R_L/20$ ohms;

R4 $\geqslant R_L$ ohms and $\geqslant 10 V_s/I_{arc}$ ohms;

R5 $\approx$ 100 k ohms;

All the arc-suppression components should be mounted close to the contacts, with all leads kept as short as possible.

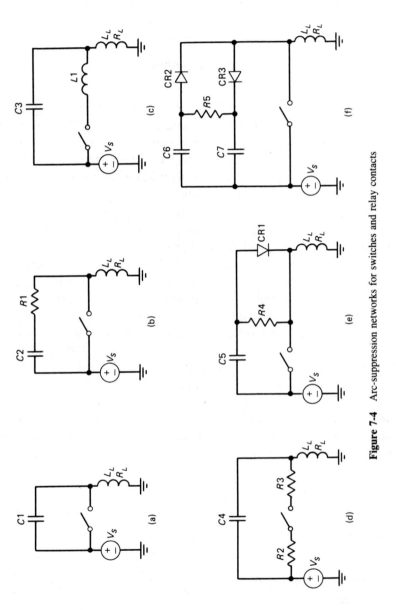

Figure 7-4 Arc-suppression networks for switches and relay contacts

Figure 7-4(b) shows the best arc suppressor for mercury-wetted switches and relays. The recommended values for $C2$ and $R1$ are

$$C2 = 10^{-7}I^2 \text{ farads, minimum of 1000 pF};$$
$$R1 = \frac{V_s}{10I^{1 + (50/V_s)}} \text{ ohms, minimum of 0.5 ohm.}$$

Contacts usually bounce on closing and can open momentarily under vibration or shock. For small relays, contact bounce usually lasts 10 to 60 μs. For power relays, contact bounce may last several milliseconds. Typical switch contacts will bounce for 5 to 50 ms. Figure 7-5 shows two circuits to keep digital circuits from seeing contact bounce. Choosing $R1C1 \approx 10^{-6}$ ohm-farads usually works well.

Spike-suppression circuits reduce the noise produced by inductive loads and eliminate the need for arc suppressors on contacts. If an inductive load is energized, and the power is suddenly cut off, the load inductance will try to maintain the current flow. If the circuit lacks arc or spike suppressors, this current charges the load's distributed capacitance. The result is a negative turn-off spike rising to ≈100 times the supply voltage in 3 μs, then falling at a rate determined by the load's resistance, capacitance, and inductance. These spikes can weld contacts, destroy transistors, and generate noise from DC to over 300 MHz.

Figure 7-6 shows five spike-suppression circuits for loads driven from DC supplies. These circuits should be mounted as close as possible to the loads to keep the current loops small and thus minimize noise problems. A simple diode across the load (CR1 in Figure 7-6(a)) limits the voltage spike to ≈1 V but extends the drop-out time of relays. By adding a few more parts (Figure 7-6(b), (c), and (e)) we can approach

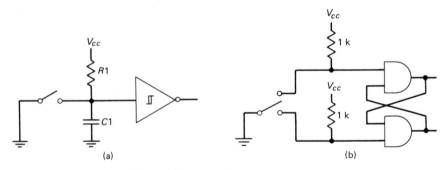

Figure 7-5   Eliminating contact bounce

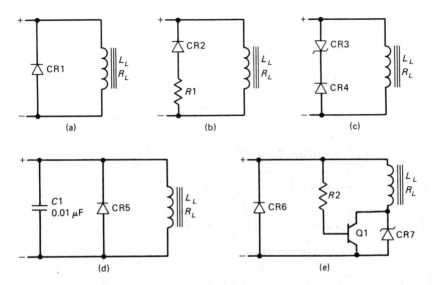

Figure 7-6   Spike-suppression networks for inductive DC loads

the drop-out time of an unsuppressed relay while limiting the turn-off spike to a safe value. Zener diode CR3 should be rated for 1.2 times the supply voltage, with a continuous current rating equal to the peak load current. In Figure 7-6(d) we add capacitor $C1$ to reduce radiated and conducted noise. If the loads are cycled intermittently, diodes CR1, CR2, CR4, CR5, CR6, and CR7 should have a continuous current rating of one-half the load current. If the loads are cycled continuously, these diodes should be rated for the full load current.

Figure 7-7 shows spike-suppression circuits for loads driven from AC or DC supplies. In Figure 7-7(a), letting $R1 \approx R_L$ ohms holds the turn-off spike to twice the supply voltage but wastes power. In Figure 7-7(b), the varistor should carry about one-tenth the nominal load current in order to limit the turn-off spike to approximately twice the supply voltage. In Figure 7-7(c), CR1 and CR2 should zener at about 1.2 times the peak supply voltage and have a continuous current rating equal to the load current. In Figure 7-7(d), $R_L/4$ ohms $\leqslant R3 \leqslant R_L/2$ ohms, and $C1 \approx L_L/(R_L R3)$ farads, rated for ten times the peak supply voltage. In Figure 7-7(e), if the peak load current is I amperes, $C2 \geqslant 10^{-6} I$ farads and $\geqslant (I/300)^2 L_L$ farads, rated for ten times the peak supply voltage, and $R4 \approx (5L_L/C2)^{1/2}$ ohms.

DC and universal motors generate substantial noise up to 20 MHz and small amounts of noise up to 1 GHz. A clean, symmetrical motor

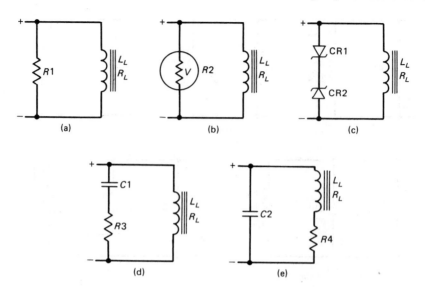

Figure 7-7   Spike-suppression networks for inductive AC or DC loads

design with sturdy brush holders, special compensation windings, and interpoles helps to reduce this noise. Motors operated at 50 V and higher should use carbon brushes with 1.5 to 2.5 mΩ resistance and a current density of 85 to 100 kA/m². Motors operated below 50 V should use low-resistance metal-graphite brushes, operated at 100 to 140 kA/m² current density. Special laminated brushes, with low-resistivity leading edges and high-resistivity trailing edges, can greatly reduce arcing at the commutator. Copper commutators build up a coating of copper oxide, making the cathode brush ten times noisier than the anode brush. Plating the commutator with ≥25 μm of chromium will reduce the cathode brush noise and extend the life of the motor.

Figure 7-8 shows spike-suppression circuits for DC motors, universal motors, and DC generators. Putting the capacitors between the field coils and the brushes (Figure 7-8(b)) allows much smaller capacitors to be used for the same amount of noise suppression. C2, C3, C4, and C5 should be bonded directly to the motor housing and should be rated for at least twice the peak supply voltage.

AC motors and alternators produce some low-frequency noise. A symmetrical mechanical and electrical design will nearly eliminate the even harmonics. Using the delta-connection instead of a wye-connection will reduce the 3rd, 6th, 9th, etc. harmonics.

My first encounter with audio rectification was about 1972, when

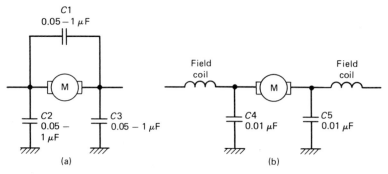

Figure 7-8   Spike-suppression networks for motors

a neighbor bought a high-power Citizens' Band transmitter. We heard about it right away, because his voice came blaring out from our electronic organ. After a couple of days we got tired of listening to his blather. My father suspected that the speaker wires were picking up the CB transmissions, so he mounted two 0.01 μF ceramic capacitors between the speaker terminals and the (grounded) amplifier chassis. That simple fix took care of the problem. (*Note:* If the capacitors are too large, the amplifier may oscillate and burn out the speaker coils; if you have access to an oscilloscope, check the amplifier outputs for spurious oscillations after applying this fix.)

RECOMMENDED READING

Consumer Electronics Systems Technician Interference Handbook—Audio Rectification. Washington, D.C.: Consumer Electronics Group/Electronic Industries Association, no date.

Motorola Thyristor Data. Phoenix, AZ: Motorola Semiconductor Products, 1985.

HOLM, RAGNAR, *Electric Contacts*, 4th ed. New York: Springer-Verlag New York Inc., 1967.

National Association of Relay Manufacturers, *Engineers' Relay Handbook*, 2nd ed. New York: Hayden Book Co., Inc., 1969.

OTT, HENRY W., *Noise Reduction Techniques in Electronic Systems*. New York: John Wiley & Sons, 1976.

8

DESIGNING POWER SUPPLIES

Switching power supplies are major noise producers, dominating conducted- and emitted-noise emissions below 30 MHz. Linear power supplies may "motorboat" (generate low-frequency sawtooths) when operated at low output currents, or oscillate when forced to drive poorly bypassed loads through long cables. Another problem is poor input-output isolation, which lets power-line noise enter the product through the power supplies and lets product noise escape. These problems can be reduced through the proper choice of components, careful layout, good bypassing, filtering, and shielding.

Figure 8-1 shows a basic linear power supply. Transformer T1 steps up or steps down the line voltage and provides isolation from the primary AC power. CR1, CR2, CR3, and CR4 rectify the secondary voltage and charge bulk filter capacitor C1. V_{ref} drives the base of Q1 to regulate the output voltage $V_{out} = V_{ref} - V_{be}$ (Q1) volts. Figure 8-2 shows the high-frequency model for this power supply—from HI to V_{out}, and from LO to V_{out}, the supply has ≈ 13 pF series capacitance and ≈ 50 pF shunt capacitance, so about 20% of the power-line noise will reach the product, and vice versa.

High-frequency noise is extremely common on most AC power lines. The average home sees 200 V spikes every few minutes, about one 400 V spike per day, and about one 1000 V spike each year. Homes in areas of high lightning activity see about two 1000 V spikes per day and about one 5000 V spike per year. Offices and manufacturing plants

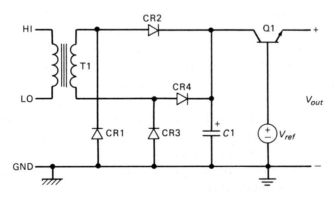

Figure 8-1   Basic linear power supply

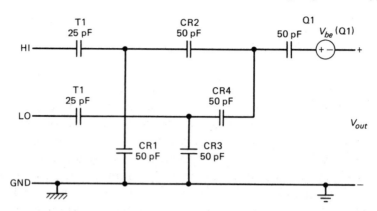

Figure 8-2   High-frequency model of basic linear power supply

tend to have much more noise, with electric motors generating 1500 V to 2500 V spikes. Lightning can put 10 to 20 kV spikes on outdoor power lines, and 2 to 6 kV spikes on indoor power lines (limited by the ≈6 kV arc-over voltage of wall sockets). As a general rule, power supplies should withstand 3 kV noise spikes without damage, and high-reliability power supplies should safely withstand 6 kV noise spikes.

The automotive environment is not much better. Load dumps, where the alternator load suddenly drops, can generate +120 V spikes. The switching of inductive loads can generate spikes between −300 V and +80 V. Coupling between wires in harnesses can generate +200 V spikes, and just turning off the ignition can generate −100 V spikes! Under normal conditions accessories produce about ±1.5 V noise and the ignition system produces ±3 V noise—which can reach ±75 V if the battery is disconnected.

One way to protect electronic equipment from voltage spikes on power lines is to use surge protectors. For example, I have my home computer, printer, color monitor, and plotter all plugged into a six-outlet "power surge controller." My system would probably work just as well without it, but I like the additional protection it provides. This particular unit contains metal-oxide varistors (MOVs), which are very nonlinear resistors. Some other types of surge protectors are silicon-carbide varistors, zener diodes, Transzorbs, Surgectors, and gas-tube arrestors. Table 8-1 summarizes the important properties of these devices.

The rectifiers in the power supply generate voltage spikes when they turn on and current spikes when they turn off. These spikes can be reduced by using soft-recovery rectifiers or rectifiers with high voltage and current

TABLE 8-1 Spike-Suppression Properties of Surge Protectors

Device Type	Breakdown Voltage (V)	Nonlinearity a*	Current Rating (A)	Capacitance (pF)	Response Time (ns)
Gas-tube arrestor	70–40,000	—	1000–10,000	0.5–10	50–5000
MOV	6–4700	15–30	10–70,000	10–33,000	<1–50
SiC varistor	9–1000	2–7	1–1000	30–4000	300–10,000
Surgector/ThyZorb	5–600	—	5–3500	90–200	10–10,000
Transzorb	6–500	≈35	2–2000	10–90,000	<1–10
Zener diode	1–700	30–100	1–500	2–60,000	<1–25

*$I = KV^a$ amperes.

ratings. Other possibilities are to limit the current through the rectifiers (Figure 8-3(a)), lower the rate at which the current changes (Figure 8-3(b) and (c)), or absorb the spikes with high-quality bypass capacitors (Figure 8-3(d) and (e)). Schottky diodes may require RC-snubbers (Figure 8-3(f)) to prevent ringing at turn-off. The spike suppressors in Figure 8-3(a), (b), (c), and (e) also block external noise, or shunt it to ground, increasing the input-output isolation of the power supply and making the product less noise-sensitive. (*Note:* The free-wheeling diode in a switching power supply must turn off faster than the switching transistor, or the power supply will self-destruct from its own noise spikes.)

Let's see what we can do to noise-proof the power supply in Figure 8-1. Our biggest problem is power-line noise, because it can damage the product. Figure 8-4(a) shows a typical commercial power-line filter. $L1$ and $L2$ block high-frequency noise, $C2$ and $C3$ bypass high-frequency differential noise, and $C4$ and $C5$ bypass high-frequency common-mode noise. Typical values are $L1 = L2 = 1.8$ to 47 mH, $C2 = C3 = 0.1$ to 2 μF, and $C4 = C5 = 0.0022$ to 0.033 μF. Higher values for $C4$ and $C5$ would improve the noise rejection, but federal safety regulations limit the ground current to 3.5 mA for grounded systems and 0.5 mA for ungrounded systems. We may also need a bleeder resistor, with resistance $R1 < 0.4\Omega F/(C2 + C3)$ to discharge $C2$ and $C3$. (*Note:* When choosing a power-line filter for a switching power supply, the filter's resonant frequency must be *less than* the switching frequency.)

Sometimes we need just a little bit of filtering, so the circuit in Figure 8-4(b) may do the job. $C3$ is a paper or plastic capacitor, and $C4$ and $C5$ are standard 1.4 kV disc ceramics. $R2$ and $R3$ are carbon-composition resistors. This circuit can be built on a terminal block or a printed circuit

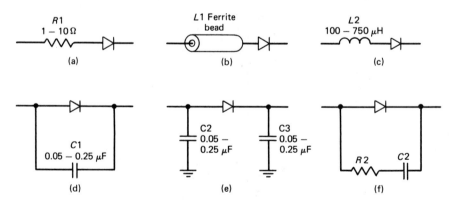

Figure 8-3   Reducing rectifier turn-on/turn-off transients

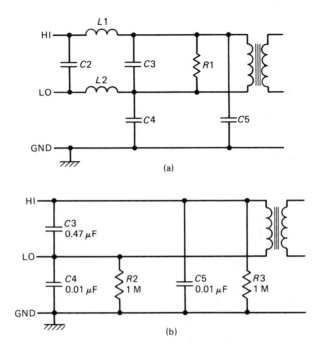

Figure 8-4 Reducing noise at the primary side

board, being careful to keep the capacitor leads as short as possible. All components used in power-line filters should be rated to withstand two times the filter's rated AC voltage and two times the rated AC current. For linear power supplies, the cut-off frequency of the filter should be at least 1.5 times the maximum AC input frequency.

We can also shield the power transformer. Regular transformers have 10 to 50 pF interwinding capacitance. A Faraday-shielded transformer with the shield attached to DC ground (Figure 8-5(a)) has ≈0.01 pF interwinding capacitance. For a double-shielded transformer, the primary's shield should be attached to earth ground and the secondary's shield to DC ground. To see if a Faraday-shielded transformer would fix a noise problem, we can hook up a center-tapped isolation transformer as shown in Figure 8-5(b). If we only need a small reduction in the interwinding capacitance, we can also try split-bobbin transformers and toroidal transformers.

Now let's see what we can do on the secondary side of the power supply (Figure 8-6). We can put ferrite beads (L3 and L4) on the transformer leads to block noise spikes, slow down charging-current pulses,

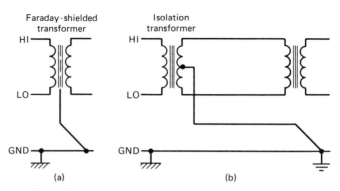

Figure 8-5 Reducing transformer-coupled noise

and reduce rectifier turn-off spikes. We can add varistor $R1$ to clamp high-voltage spikes, and (small) capacitors $C6$ and $C7$ to shunt high-frequency noise to ground. To keep product noise from entering the power supply we can add feedthrough capacitor $C8$ and ferrite bead $L5$. (*Note: $L3$ and $L4$ are most effective with low-impedance loads.*)

Figure 8-7 shows the high-frequency model of the fully noise-suppressed power supply. $L1$, $L2$, $L3$, $L4$, and $L5$ block high-frequency noise trying to travel between the AC power lines and the product. $R1$ clamps high-voltage spikes, and $C2$, $C3$, $C4$, $C5$, $C6$, $C7$, and $C8$ short out high-frequency noise. Very little noise will get through the power-supply circuitry, but we must keep the AC wiring well away from the DC wiring to preserve the input-output isolation.

Switching power supplies may emit excessive noise because of capacitive coupling between the switching transistor(s) and the heatsink. A TO-3 transistor on a mica insulator will have 100 to 250 pF parasitic capacitance. This capacitance can be reduced to ≈ 1 pF by placing a

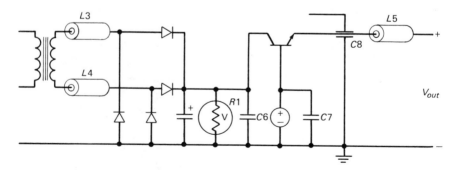

Figure 8-6 Reducing noise at the secondary side

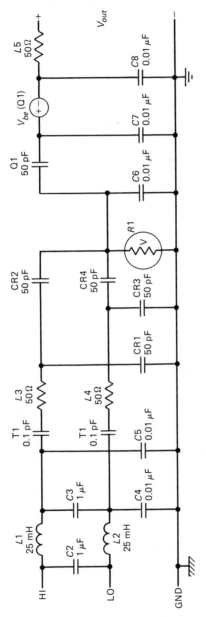

Figure 8-7 High-frequency model of noise-suppressed linear power supply

shield between the transistor and the heatsink and connecting the shield to DC ground. Bergquist manufactures special insulators, called "Sil-Pad Shields," that are designed just for this purpose.

Consider using separate power supplies for high- and low-frequency circuits. Similarly, if a product has high- and low-power circuits, the low-power circuits should have their own power supply or a voltage regulator for isolation. If a power supply does not have remote sensing, one of the power-supply terminals should be connected to chassis ground. If a power supply with remote sensing is driving a single load, one of the load terminals should be connected to chassis ground. If a power supply with remote sensing is driving multiple loads, connect one of the sense points to chassis ground.

In 1980 I was developing a tester for a power supply and found the supply's $+5$ V output varying between $+5.1$ V and $+6.5$ V. Observing the $+5$ V output with an oscilloscope, I saw a sawtooth waveform falling to $+5.1$ V over 100–200 ms, then jumping to 6.0–6.5 V in ≈ 50 µs. The apparent scenario was as follows: (1) output voltage is too high, so the pass transistor is turned off; (2) the output voltage gradually falls to $+5.1$ V, and the pass transistor starts to turn on; (3) the output immediately charges the output capacitance, overshoots the desired voltage, and turns the pass transistor off again. When I checked with the power supply's designer, he told me that the $+5$ V output was designed for a minimum load of 100 mA and 200 µF. So I added a 220 µF capacitor and a 50 Ω resistor to the tester. That tester has now tested well over 100,000 power supplies without another noise problem.

RECOMMENDED READING

1984 SAE Handbook, Vol. 2, Parts and Components. Warrendale, PA: Society of Automotive Engineers, Inc., 1984.

CHERNIAK, STEVE, "A Review of Transients and Their Means of Suppression." Phoenix, AZ: Motorola Semiconductor Products Application Note AN-843.

GINSBERG, GERALD L., *A User's Guide to Selecting Electronic Components.* New York: John Wiley & Sons, 1981.

GOTTLIEB, IRVING M., *Regulated Power Supplies.* Indianapolis: Howard W. Sams & Co., Inc., 1978.

MARDIGUIAN, MICHEL, *Interference Control in Computers and Microprocessor-Based Equipment.* Gainesville, VA: Don White Consultants, Inc., 1984.

9

PARTITIONING

Partitioning is the process of deciding *what* should go *where* in an electronic system. We can minimize noise and interference problems by isolating the sensitive circuits from the noisy ones through control of common-impedance coupling, inductive coupling, capacitive coupling, and antenna effects. We need to (1) keep low-power (sensitive) circuits close to the signal sources, (2) keep high-power (noisy) circuits close to the loads. (3) separate low-power and high-power circuits as much as possible, (4) keep wires as short as possible, and (5) keep current loops as small as possible.

We start by sorting the circuits into five groups: (1) sensitive high-impedance circuits ($| Z | \geq 376.7 \ \Omega$, subject to capacitive coupling), (2) sensitive low-impedance circuits ($| Z | < 376.7 \ \Omega$, subject to inductive coupling), (3) medium-sensitivity/medium-power circuits, (4) high-voltage circuits, and (5) high-current circuits. Analog circuits usually fall into the first two groups, digital circuits into the third group, and interface circuits and power supplies into the last two groups. We can safely combine circuits within a group to form subassemblies, but we must keep high-impedance circuits well away from high-voltage circuits, and low-impedance circuits well away from high-current circuits. In general, when we must connect circuits in different groups, the interconnecting signals should have medium sensitivity and medium power.

The circuits within a given subassembly should have similar input/output requirements and similar noise characteristics. This requirement divides most systems into analog, digital, power-supply, and electromagnetic device/driver subassemblies. We may also want to put high- and low-frequency circuits in separate subassemblies. Each subassembly should be as small and compact as possible, with its own power/ground system. The wiring within the subassemblies should be designed for low impedance, keeping current loops as small as possible.

Mount transformers, solenoids, and other electromagnetic devices so that their magnetic fields are perpendicular to one another and well away from cables. Design cables for minimum length, minimum impedance, and minimum loop area. Cables for high-speed logic should have at least one-fifth of the conductors allocated as grounds; cables for medium- or low-speed logic should have at least one-tenth of the conductors allocated as grounds. Keep sensitive circuits and their cables away from the other circuits, using the system's support structure to provide free shielding. Tie the subassembly ground systems together at only one point, and put grounded shields around high-impedance circuits and high-voltage circuits.

10

GROUNDING

Every grounding system is a compromise between conflicting requirements. Ground systems must:

form voltage-reference networks for signals (typically ± 100 mV for analog circuits, ± 200 mV for digital circuits)

carry signal-return currents,

carry power-return currents,

form reference planes for antennas,

keep high-frequency potentials from developing near antennas,

protect people and equipment from lightning,

protect people and equipment from power-line faults, and

bleed off static charges.

A grounding system must be carefully designed to meet all these requirements, while minimizing the unwanted coupling between signals that causes noise problems.

The January–March 1983 issue of *EMC Technology* magazine contains five excellent articles on grounding; on page 44 Henry Ott defines ground as *"a low-impedance path for current to return to the source."* Using this definition, we can see that *any* current flow in a ground system will cause voltage differences. If we want equipment to work properly, these voltage differences must be insignificant compared to the signals using the grounds. So our goals in designing a ground system are (1) *to keep ground impedances low* and (2) *to control the current flow between sources and loads.*

Our first question must be: "How big is the system?" If we apply a signal with frequency f (hertz) and wavelength $\lambda \approx 2.998 \times 10^8/f$ meters to an l-meter-long piece of a conductor, we would expect to see $|Z|$ ohms impedance (Appendix D). But *standing waves* increase the effective impedance by $\tan(2\pi l/\lambda)$, so we actually see $|Z|[1 + \tan(2\pi l/\lambda)]$ ohms impedance. If two points are $\lambda/4$, $3\lambda/4$, $5\lambda/4$, $7\lambda/4$, . . . meters apart, the conductor acts like an open circuit.

Therefore, to keep voltage differences low, we must limit the size of the grounding system. For military equipment, transmitters, receivers, and other sensitive systems, the maximum distance between grounds

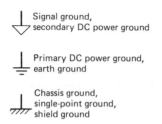

Signal ground,
secondary DC power ground

Primary DC power ground,
earth ground

Chassis ground,
single-point ground,
shield ground

Figure 10-1   Standard ground symbols

should be ≤ 0.05λ, where λ is the wavelength of the highest-frequency signal used in the system. This limits the ground impedance to 133% of its nominal value. Most commercial systems can stand 0.10λ spacing between grounds, limiting the ground impedance to 173% of nominal. Insensitive systems can stand 0.15λ spacing between grounds, for ground impedances up to 238% of nominal.

We can isolate signal-return, DC-power-return, and AC-power-return currents from one another by building the system with three independent ground networks (Figure 10-1), tied together at only one point. This lets us optimize each ground system without worrying about the others. Signal grounds, for example, must have low impedance from DC to megahertz but don't carry much current. DC-power grounds need low impedance from DC to kilohertz but must carry much higher currents, while AC-power/chassis grounds need low impedance in the 100 Hz range and may carry hundreds of amperes (a typical requirement is ≤100 mΩ resistance and ≤100 μH inductance, requiring #12 copper wire or #10 aluminum wire).

Figure 10-2 shows a floating-ground system for extremely sensitive circuits. This setup requires complete isolation between the circuits and the chassis—high resistance and low capacitance—and is very touchy. The circuits must be powered by solar cells or batteries, and signals must enter and leave through transformers or optoisolators. To prevent static buildup, some designers put a high-value bleeder resistor between the signal ground and chassis ground.

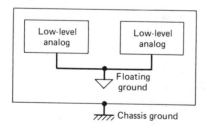

Low-level analog

Low-level analog

Floating ground

Chassis ground

Figure 10-2   Floating-ground system

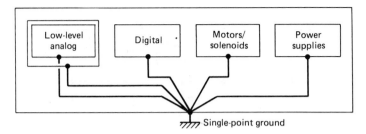

Figure 10-3   Single-point-ground system

Figure 10-3 shows a pure single-point-ground system. Each circuit and each shield has its own connection to the single-point ground. Each rack and each frame has *one* bond to the chassis. This setup eliminates common-impedance coupling and low-frequency ground loops. Single-point grounds work very well up to 1 MHz and can be used up to 10 MHz in small systems (maximum dimension under 0.05λ). But sensitive analog circuits can still pick up inductive- and capacitive-coupled noise, despite the numerous ground wires. Nevertheless, most military and aerospace applications require pure single-point-ground systems.

Figure 10-4 shows a modified single-point-ground system. Circuits with similar noise characteristics are tied together, with the most-sensitive circuits closest to the single-point ground. This setup reduces the total number of ground wires needed, at a slight increase in common-impedance coupling. When a card has separate analog and digital grounds, these

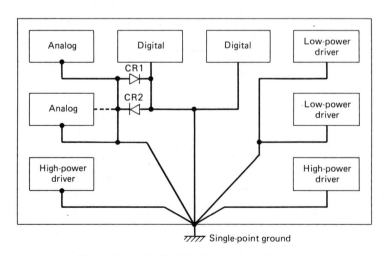

Figure 10-4   Modified single-point-ground system

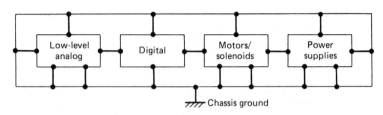

Figure 10-5 Multipoint-ground system

should be connected together by back-to-back diodes (CR1 and CR2 in Figure 10-4) to protect the card from static buildup when it is not installed in the system.

Figure 10-5 shows a multipoint-ground system. The circuits and chassis sections are bonded together with numerous short ($l < 0.1\lambda$) jumpers to minimize standing waves. We usually use this setup for high-frequency circuits ($f \geq 10$ MHz) with similar noise characteristics. This setup requires careful maintenance, produces many ground loops, and should not be used for sensitive circuits.

Figure 10-6 shows hybrid-ground systems formed by combining floating-, single-point-, and multipoint-ground systems. Figure 10-6(a)

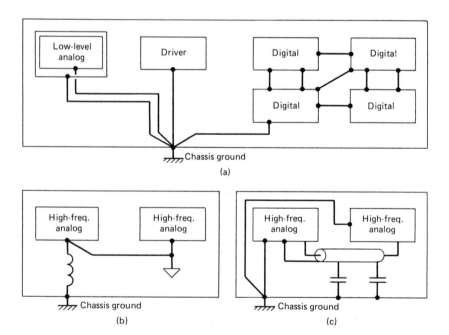

Figure 10-6 Hybrid-ground systems

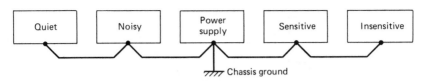

Figure 10-7   Daisy-chained-ground system

shows a single-point-ground system with multipoint grounding for digital logic (a very common arrangement). In Figure 10-6(b), an ≈1 mH inductor bleeds off static while isolating the high-frequency circuits from chassis ground. In Figure 10-6(c), capacitors every 0.1λ along an insulated cable prevent high-frequency standing waves while avoiding low-frequency ground loops. When we use these last two arrangements, we must be very careful about resonances in the grounding system caused by the system's parasitic capacitance and inductance.

Daisy-chained grounds (Figure 10-7) are very common and extremely troublesome. When we must use this arrangement, we can minimize the noise problems by putting noisy circuits on one daisy chain and sensitive circuits on another. Circuits on the "noisy" daisy chain should be in the order noisy–noisier–noisiest–chassis ground. Circuits on the "quiet" daisy-chain should be in the order sensitive–more sensitive–most sensitive–chassis ground.

Sensitive analog circuits require tight control over ground currents. Figure 10-8(a) shows an opamp circuit using the DC-power ground as the signal return. This circuit will pick up noise through common-impedance coupling and inductive coupling. By adding a signal return and two bypass capacitors, as in Figure 10-8(b), we shrink the signal loop area to reduce inductive coupling, and we separate the signal-return current and the DC-power return current to reduce common-impedance coupling.

Digital circuits are insensitive to low-frequency noise but need grounds with low impedance at high frequencies. Transmission lines meet this requirement quite nicely (Appendix E). Twisted-pair and coaxial cables provide one signal return per signal, while flat cables should use at least every fifth or tenth conductor as signal returns. These signal returns should be tied to ground at the drivers and receivers.

Figure 10-9(a) shows a digital circuit with twisted-pair cable 25 mm above a groundplane. The twisted-pair cable forms a 105 Ω transmission line, while the signal conductor and the groundplane form a 468 Ω transmission line. As a result (Figure 10-9(b)), 82% of the signal current returns through the cable. We have effectively isolated the signal-return

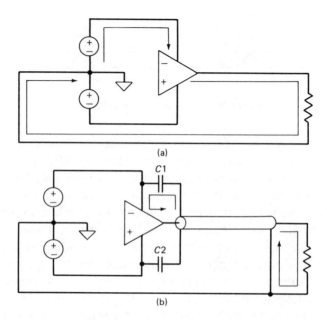

Figure 10-8   Use of bypass capacitors and dedicated returns to shrink current loops

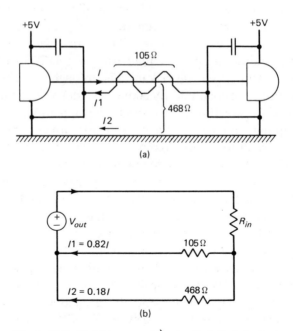

Figure 10-9   Shrinking ground loops with dedicated returns

current from current flowing in the groundplane and greatly reduced the common-impedance noise and inductively coupled noise in the circuit.

Digital circuit boards should have groundplanes or gridded ground nets (Figure 10-10) to provide low-impedance signal grounds at high frequencies. Up to 40% of a groundplane may be used for signal and power wiring without seriously increasing the ground impedance. Ground lands and power lands should be as wide as possible, but even narrow ground lands will help reduce the ground impedance.

If we take two conductors, with inductances $L1$ and $L2$ and mutual inductance M, and connect them in parallel, the overall inductance becomes

$$L = \frac{L1L2 - M^2}{L1 + L2 - 2M} \text{ henries.}$$

If the conductors are separated by at least three times their width (diameter), $M \ll L1$ and $M \ll L2$, and the inductance (and the high-frequency impedance) is chopped roughly in half. In essence, at high frequencies the two conductors act like a single conductor spanning both conductors. For example, if we parallel two 0.254-mm-wide lands that are 7.68 mm apart, we get approximately the same impedance as a single 8.188-mm-wide land. Because of this effect, a gridded ground net has approximately the same impedance as a solid groundplane.

Shields must be grounded to block electric fields. Shields should be grounded close to signal entry/exit points (Figure 10-11(a)) to minimize current flow in the shields. Shields around sensitive circuits should be grounded at only one point in order to prevent ground loops. Cable shields should be grounded at least every 0.2λ to prevent standing waves.

Chassis grounds should have low impedance at high currents and should be easy to maintain. Jumpers connecting panels to the chassis

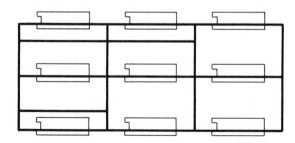

Figure 10-10   Gridded ground net

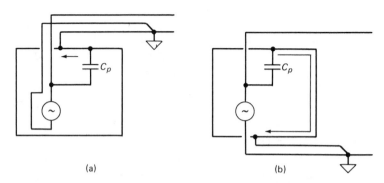

Figure 10-11   Positions for shield grounds: (a) proper, (b) improper

should be less than 0.1λ long and less than 0.1λ apart (maximum of 0.2 meter). Keep chassis-ground jumpers well away from sensitive circuits.

Ground loops pick up noise through inductive coupling and are usually easiest to break at the load. If many sources drive one load, break the ground loops at the sources. Figure 10-12 shows three ways to break ground loops. Transformers (Figure 10-12(a)) give good isolation up to 5–10 MHz. For better isolation, use a Faraday-shielded transformer with the shield grounded at the load. Common-mode chokes (Figure 10-12(b)) give good isolation for frequencies above $5R/(2\pi L)$ hertz (R in ohms, L in henries). An easy way to make a common-mode choke is to run the cable through a toroid. Optoisolators (Figure 10-12(c)) have very good performance from DC into the MHz range. Standard optoisolators are

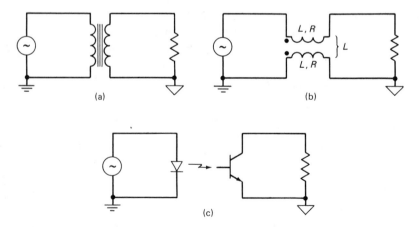

Figure 10-12   Breaking groundloops: (a) transformer, (b) common-mode choke, (c) opto-isolator

rated for 1.5 kV to 7 kV common-mode voltage, with under 2 pF coupling capacitance, while shielded optoisolators have less than 0.5 pF coupling capacitance. For even better isolation, use LEDs and photodetectors coupled by fiber-optic cables.

In early June 1983 Ted Hall asked me to look at a stepper-motor tester. The designer had spent the last two weeks battling severe noise problems and had asked for help. The problem seemed to be a servo-amplifier that produced 24 V, 3 A pulses at roughly 16 kHz. These pulses put $\approx$2 V spikes on the signal, power, and ground lines and drove the microcomputer and the other electronic circuits absolutely nuts.

According to the cable drawings, all the ground wires were in the power cables. When we examined the tester, I saw that power cables and signal cables were bundled together and frequently formed loops with 0.5 to 1 m^2 area. We spent a total of four hours looking at the tester documentation and the tester, discussing possible solutions. After eliminating all the easy solutions, I suggested ripping out all the power and signal cabling and rewiring the tester following these rules:

1. Designate a terminal block near the power supplies as a single-point ground.
2. Run a separate ground wire from this terminal block to each power supply and card cage, twisted with the power wires to that unit.
3. Use twisted-pair cables for the signal cables, with every other wire connected to ground at both ends (digital ICs ignore ground-loop noise).
4. Use Schmitt-trigger gates on all incoming signals.

After working two days to rewire it, we tried the tester again. Everything worked beautifully, except for a gas-discharge display that flickered when the stepper motor was running. We added a Mu-Metal shield around the motor to cure that problem. This tester has been in daily use for three years now, testing many tens of thousands of motors, without any noise problems.

RECOMMENDED READING

DC Power Supply Handbook. Berkeley Heights, NJ: Hewlett-Packard, 1970.
Interference Reduction Guide for Design Engineers, Vol. 1. Springfield, VA: NTIS (AD 619 666), 1964.

DENNY, HUGH W., *Grounding for the Control of EMI*. Gåinesville, VA: Don White Consultants, Inc., 1983.

EVERETT, WOODROW W., JR., *Topics in Intersystem Electromagnetic Compatibility*. New York: Holt, Rinehart and Winston, Inc., 1972.

OTT, HENRY, "Ground—A Path for Current Flow," *EMC Technology*, 2:1 (January–March 1983), 44–48.

11

BONDING

Bonding is the joining of conductors to form stable, homogeneous current paths with negligible resistance. Good bonds have ≤0.5 mΩ resistance and ≤25 nH inductance, while bonds for radio-frequency currents should have ≤80 mΩ impedance up to 20 MHz. The best bonds are direct, permanent, homogeneous metal-to-metal contacts formed by welding, brazing, exothermic bonding (such as Cadweld^{tm}), silver-soldering, or soldering, with seam lengths greater than the conductor overlap. Grounding systems for lightning protection should have at least #12 copper wire or #10 aluminum wire conductors with ≥5 mm^2 bond cross sections. (*Note:* Soft-soldered bonds are not permitted in conductors that may carry lightning currents.)

Second best are direct metal-to-metal compression bonds. Prepare the bonding surfaces of the conductors by

1. cleaning an area about 50 percent larger than the final bond with a wire brush, steel wool, or abrasives (7/0 garnet paper works very well for this);
2. wiping off the debris;
3. cleaning the surfaces with solvent; and
4. thoroughly drying the surfaces with a clean cloth.

The bonding surfaces should be clean and bright after this treatment. Within one hour after cleaning the surfaces, clamp the conductors together with bolts, rivets, or machine screws to apply 8300 to 10,300 kPa pressure on the joint (1200 to 1500 psi, the lower pressure for soft metals). Star washers or lock washers under the bolt heads and nuts will help keep the joint tight. A bond with ≥650 mm^2 cross section will have less than 0.1 mΩ resistance. Table 11-1 gives the maximum current ratings and the required clamping torque for bonds using machine screws or bolts.

Third best are compression bonds between metal surfaces with conductive coatings. These bonds typically have a few milliohms resistance, but over one ohm impedance at 1 MHz. The metals may be plated with cadmium, tin, or silver or coated with Alodine #1000, Dow #1, Dow #15, Iridite #14, Iridite #18p, or Oakite #36.

Conductive pastes can be used to seal seams in screen rooms and other places where conductors may shift slightly. Carbon-loaded epoxy

TABLE 11-1   Maximum Current Capacity and Required Clamping Torque
for Direct Metal-to-Metal Bonds

Bolt Size	Maximum Current		Clamping Torque*
	1 Wire	≥2 Wires	
M3 × 0.5	10 A	7 A	0.56–0.74 N-m
#5–40	11 A	8 A	0.78–1.05 N-m
#6–32	12 A	9 A	1.1–1.5 N-m
M4 × 0.7	19 A	14 A	1.4–2.0 N-m
#8–32	22 A	15 A	1.8–2.5 N-m
#10–32	32 A	22 A	2.6–3.6 N-m
M5 × 0.8	36 A	25 A	2.7–3.8 N-m
#12–32	46 A	33 A	3.4–4.8 N-m
M6 × 1	53 A	37 A	5.0–7.0 N-m
1/4–20	56 A	40 A	6.6–9.5 N-m
M7 × 1	83 A	58 A	7.4–11 N-m
5/16–18	100 A	70 A	12–18 N-m
M8 × 1	120 A	84 A	10–15 N-m
3/8–16	170 A	119 A	22–32 N-m
M10 × 1.25	210 A	147 A	21–31 N-m

*1 N-m ≈ 8.85 inch-pounds.

has about 0.1 Ω-m resistivity. Silver- and gold-loaded epoxies have 10 to 57,000 nΩ-m resistivity, with 60%–70% loading (by weight) giving the best combination of adhesion and resistivity. These bonds should be clamped with about 69 kPa (10 psi) pressure.

To bond removable or shock-mounted assemblies, use solid metal straps, braided straps, or wires. Place straps and jumpers where they are easy to inspect but protected from accidental damage (while still meeting the spacing requirements). Straps should be short and broad, with a length-to-width ratio under 5, and preferably under 3, to minimize ground imped-ance. *Grounding straps should never be connected in series.* Solid copper straps should be ≤0.1 mm thick and ≥30 mm wide, while solid aluminum straps should be ≤0.2 mm thick. Braided straps are more flexible than solid straps, but they corrode easily and they may fray (and at high frequencies, the broken wires may act like antennas). Ground wires should be at least 18 AWG, with crimped-on terminals. "Fast-on" terminals work well for wires that must be disconnected frequently. External-tooth star washers under the terminals will cut through surface debris and help form solid contacts. The resonant frequency of a strap or jumper wire should be at least 16 times the highest frequency we are trying to ground or bypass.

Do not depend on self-tapping screws, screw threads, Tinnerman nuts, bearings, hinges, or slides for bonding. They do not provide reliable contact, and the ground currents can cause severe corrosion. Shafts can be grounded by a brush riding on a slip ring, or by a phosphor-bronze finger riding on the end of the shaft. Hinges require bonding straps every 50 mm along their length. Slides can be bonded with braided straps or wires.

Rough or irregular surfaces, or joints that must be RF-tight, may need conductive gaskets. These gaskets are screwed or cemented to one of the surfaces and must be protected from damage. In general, use narrow flanges for rough surfaces and wide flanges for smooth surfaces (see Chapter 14 for details).

Within one week after forming a bond, apply a protective finish (paint, silicone rubber, grease, polysulfates) to resist moisture and gas penetration and thus prevent corrosion. Apply the finish to both members or just to the cathodic member (Appendix B). When joining unlike metals, make the anodic members larger than the cathodic members. If the metals are not in the same or adjacent groups, put an even-more-anodic jumper, washer, bolt, or clamp in the joint to protect the structural members. (*Note:* This sacrificial hardware must be replaced periodically as part of preventive maintenance on the system. Tin and cadmium platings can also help reduce corrosion.)

Most grounding straps, jumpers, and wires are made of aluminum, tinned copper, or copper and may require special precautions to minimize corrosion. Use aluminum-alloy washers or cadmium- or zinc-plated hardware for connections to group I metals (Appendix B). Make direct connections to group II and group III metals. To connect aluminum to group IV metals, use cadmium- or tin-plated washers, stainless-steel hardware, or cadmium- or zinc-plated hardware. Connect tinned copper directly to group IV metals, and connect copper directly to group IV and group V metals. Do not connect aluminum or tinned-copper straps or jumpers to group V metals.

RECOMMENDED READING

Handbook on Radio Frequency Interference, Vol. 3. Wheaton, MD: Frederick Research Corp., 1962.
Interference Reduction Guide for Design Engineers, Vol. 1. Springfield, VA: NTIS (AD 619 666), 1964.

DENNY, H. W., et al., *Grounding, Bonding, and Shielding Practices and Procedures for Electronic Equipments and Facilities*, Vol. I. Springfield, VA: NTIS (AD A022 332), 1975.

DENNY, HUGH W., *Grounding for the Control of EMI*. Gainesville, VA: Don White Consultants, Inc., 1983.

EVERETT, WOODROW W., JR., *Topics in Intersystem Electromagnetic Compatibility*. New York: Holt, Rinehart and Winston, Inc., 1972.

FICCHI, ROCCO F., ed., *Practical Design for Electromagnetic Compatibility*. New York: Hayden Book Co., Inc., 1971.

12

DESIGNING
CIRCUIT BOARDS

After partitioning the system into subassemblies and designing the ground system, we must design each subassembly. We must decide how to mount the components, how to dissipate the heat they generate, and how to connect them to power, ground, and one another. In most cases we will put everything we can on circuit boards, mounting just the heavy, bulky, and high-power components on the chassis or on heatsinks. A simple subassembly may fit on one circuit board, while a complicated subassembly may have many circuit boards plugged into a backplane or "mother board."

The three most common types of circuit boards are wirewrap boards, printed circuit boards, and Multi-wire^{tm} or stitch-wire boards. Wirewrap boards are frequently used for prototypes and limited-production items. They are easy to design and modify but are hard to mass-produce, and they take up a lot of space. Printed circuit boards are used for mass-production items, high-speed systems, and space-limited systems. Printed circuit boards are easy to assemble, compact, and have good, consistent high-frequency characteristics. Their major drawbacks are the design time required and the difficulty of making design changes. (*Note:* At least one company now offers a computer-controlled router, and design software, that carves regular copper-clad laminate to form printed circuit boards. These systems can manufacture a single- or double-sided prototype board in an hour or so.) Multi-wire^{tm} boards are used for limited-production items that need the small size and good high-frequency characteristics of printed circuit boards but can't afford the design time or design cost.

The eight major steps in designing a circuit board are

1. determine the board's size, shape, and connector positions;
2. decide where to put the circuits;
3. decide on the power/grounding scheme for the board;
4. decide where to put the components;
5. lay out the power and ground lands;
6. lay out the clock lands;
7. lay out the rest of the signal wiring; and
8. clean up the design.

The physical design of the product may determine the board's size, shape, and connector positions. If you have a choice, make boards squarish instead of long and slender, because square boards are easier to design and build, and the wires tend to be shorter. Make a rough sketch of the board, indicating reserved areas for card guides, mounting hardware, tooling holes, connectors, and any other components that must occupy fixed positions.

The second step is deciding where to put the circuits. On your sketch, start by putting input/output circuits next to their connectors, and work outward from there. Put related circuits next to one another while trying to keep sensitive circuits away from noisy circuits. At this point your layout might look like one of the boards in Figure 12-1, with fast logic circuits (clocks, bus-interface logic) next to the main connector, interface circuits next to the interface connectors, and analog circuits isolated from the digital circuits. Large RAM arrays should be split in half with the

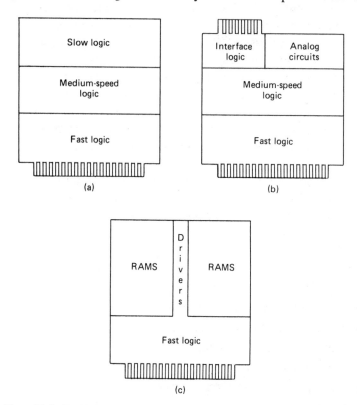

Figure 12-1 Partitioning digital printed circuit boards: (a) processor board, (b) interface board, (c) RAM board

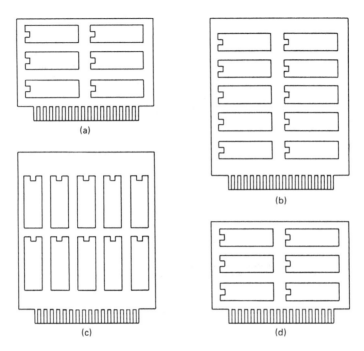

Figure 12-2 Preferred IC orientations: (a) double-sided, (b) double-sided, (c) multilayer, (d) multilayer

drivers down the middle. Also decide on the preferred orientation for integrated circuits (ICs) and other large components. For double-sided boards, the ICs should be parallel to the connectors (Figure 12-2(a) and (b)). For multilayer boards, the ICs should be parallel to the long axis of the board (Figure 12-2(c) and (d)).

The third step is deciding on the power/grounding scheme. The four options are: (1) just run the power and ground lands, and hope for the best; (2) use coplanar lines; (3) use laminar busses; or (4) use groundplanes (Figure 12-3). Coplanar lines work well on wirewrap boards and double-sided printed circuit boards (Figure 12-4). Laminar busses help reduce V_{cc}-to-ground noise but require special busbars. These busbars can run parallel to or under the ICs and can carry 2.5 to 15 amperes current. They can provide 0.001 to 2 μF/m distributed capacitance, with 14 to 35 nH/m inductance, for a power/ground impedance of 0.15 to 5 Ω. To minimize ground noise, the boards must also have large ground lands perpendicular to the bus bars.

Multilayer printed circuit boards can use solid groundplanes and voltage planes, or they can split voltage planes between several supply

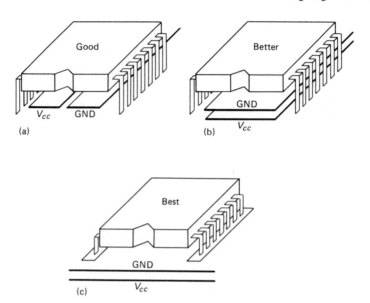

Figure 12-3   Power/ground bussing: (a) coplanar lines, (b) laminar bus, (c) ground-plane

voltages. For the best noise performance, the top and bottom layers of the board should be groundplanes or voltage planes, with signal planes in the middle. A voltage plane over a groundplane has 0.1 to 10 $\mu F/m^2$ distributed capacitance and extremely low inductance. Multi-wire[tm] boards have one groundplane and use 34 AWG wires for power and signals. In any case, sketch in the proposed power/ground system to help guide you through the next steps.

The fourth step is to position the components within each circuit. Make an accurate drawing of the board and the connectors. Using your sketch as a guide, draw in all the components whose positions are critical (microprocessors, optoisolators, isolation transformers, common-mode chokes, filters, etc.). If a component is shared by several circuits, place it near the common boundary. Place bulk capacitors and ferrite beads as close as possible to the power and ground pins on the connectors. Put bypass capacitors, decoupling networks, arc-suppression networks, and spike-suppression networks as close as possible to the components they are supposed to protect (≤ 37 mm away). (*Note:* IC sockets with integral capacitors put the bypass capacitors directly underneath the ICs and may save having to re-layout a board.)

For analog circuits, position the components so that the power and ground lands will isolate the inputs from the outputs (Figure 12-6). Leave room for feedback capacitors and bypass capacitors, which may need to be added later. Separate unshielded inductors from one another, or put them at right angles, to reduce inductive coupling. In general, try to position the components so that the signal nets will be as short and compact as possible. You will probably have to repeat this step several times before arriving at a satisfactory layout.

For wirewrap boards, that essentially finishes the design. You may need to generate wirelists and assembly drawings, or you may be able to build the board just from the layout and the schematic. When building wirewrap boards, make the power and ground connections first, then run the signal wires. For minimum power/ground inductance, trim the power and ground pins to ≈ 6 mm long, bend them into semicircles, and solder them directly to the lands on the board. Special clips that connect wirewrap pins to the lands also work well. To minimize noise in the signal wires, first run all the long wires ($\geq 1/2$ the board diagonal), going point-to-point. Then run the short wires, again going point-to-point. For added noise protection, finish by connecting the ground pins with wirewrap wires to form a ground grid over the signal wiring.

For Multi-wire^{tm} boards you need to generate a wirewrap list, an assembly drawing, and a list of coordinates for the component pins. Multi-wire^{tm} boards have a 50 μm-thick copper groundplane on one side and a web of 34 AWG polyimide-insulated copper wires (laid down by a special machine) on the other. This wiring has a nominal 55 Ω impedance, with ≈ 1 pF crossover capacitance and 2000 V breakdown voltage.

For printed circuit boards, the fifth step is to lay out the power and ground lands. Figure 12-4(a) shows a very common layout, which suffers from large current loops, high power/ground inductance, and high noise. Putting the bypass capacitors alongside the ICs (Figure 12-4(b)) shrinks the current loops, reducing inductance and V_{cc}-to-ground noise. Running power and ground lands under the ICs (Figure 12-4(c)) shrinks the current loops even more, further reducing power/ground inductance and V_{cc}-to-ground noise. Adding cross-ties (Figure 12-4(d)) creates gridded power and ground nets with small current loops, very low inductance, low V_{cc}-to-ground noise, *and low ground-to-ground noise.* Using this last arrangement, double-sided boards can approach the noise performance of expensive multilayer boards!

To get a quantitative comparison, I tested each of these power/ground layouts on a small prototype board. I populated the board with U-shaped wires to represent ICs, using 0.01 μF ceramic capacitors for

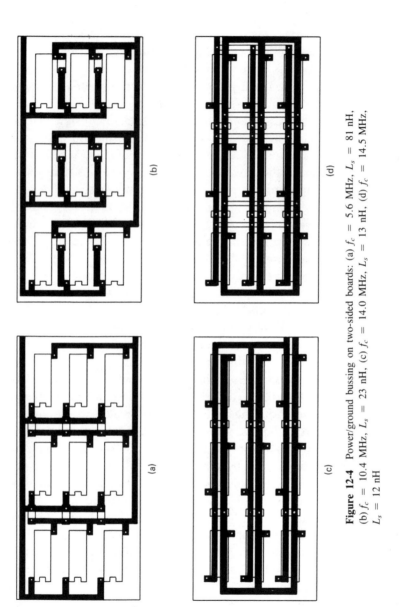

Figure 12-4 Power/ground bussing on two-sided boards: (a) f_c = 5.6 MHz, L_s = 81 nH, (b) f_c = 10.4 MHz, L_s = 23 nH, (c) f_c = 14.0 MHz, L_s = 13 nH, (d) f_c = 14.5 MHz, L_s = 12 nH

bypassing, and measured the resonant frequency of the power/ground loops with a grid-dip meter (see Chapter 2). The caption for Figure 12-4 shows the results—the resonant frequencies ranged from 5.6 MHz to 14.5 MHz, corresponding to power/ground inductances of 81 nH down to 12 nH. *The power/ground layout in Figure 12-4(d) has one-sixth the inductance, and thus one-sixth the noise, of the power/ground layout in Figure 12-4(a)!*

The sixth step is to lay out the clock wiring, keeping the clock wiring close to digital grounds and away from sensitive circuits. A good practice is to run pairs of lands, one for the clock and one for the clock return. Connect the clock return to digital ground near every IC that drives or receives the clock. For multilayer boards run clock lands next to groundplanes or voltage planes. Keep clock loops very small, because clocks and their harmonics tend to dominate the emitted radiation, and one ≥ 0.001 m^2 loop can easily exceed the FCC limit on radiated noise.

Lay out the rest of the signal lands normally. For arrays of RAM chips, run address lines in one direction and data and enable lines in the other. The land for the least significant bit of the address should be next to a ground land. Try to make lands at least 1/150th as wide as they are long, and avoid zigzag routes. Don't make sharp bends (Figure 12-5(a)) when laying out the lands. Chopping off corners, limiting bends to 45°, or using gentle curves (Figure 12-5(b), (c), (d)) will keep the land impedance nearly constant from DC to several gigahertz.

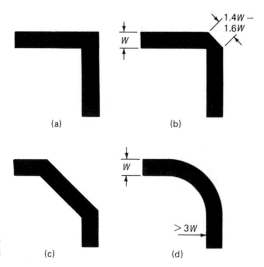

Figure 12-5   Bends in printed circuit board lands: (a) poor, (b) good, (c) good, (d) good

For analog circuits, use the power and ground lands to isolate the inputs from the outputs (Figure 12-6). Make all signal lands as short and compact as possible. Form guard rings around sensitive input signals and tie them to ground (Figure 12-6(b)) or to a low-impedance output (Figure 12-6(a)) to protect the inputs from leakage currents (we can use teflon inserts for very sensitive signals). Leave room for adding feedback capacitors and bypass capacitors to the circuits (see Chapter 5). If possible, the component side of the board should be a solid groundplane.

The final step is: (1) widen the power and ground lands as much as possible; (2) wherever there is room, run lands to tie digital grounds together; and (3) fill in open areas with copper, connected to ground through lands or vias. This forms a massive ground grid to minimize current loops, ground impedance, V_{cc}-to-ground noise, and ground-to-ground noise. (*Note:* Don't tie grounds together haphazardly—the overall grounding scheme *may require* several independent ground nets on a board for noise control.)

If a circuit board must be redesigned, test the current design to locate opportunities for improvement. Measure the ground-to-ground noise between ICs and the V_{cc}-to-ground noise at each IC. Observe clocks and other critical signals with an oscilloscope, looking for ringing at the ends of signal nets. V_{cc}-to-ground noise above $0.1V_{cc}$, too-slow rise times, and ringing on the rising edges of signals all indicate that the bypass capacitors are too small. Ground-to-ground noise above 150–300 mV indicates

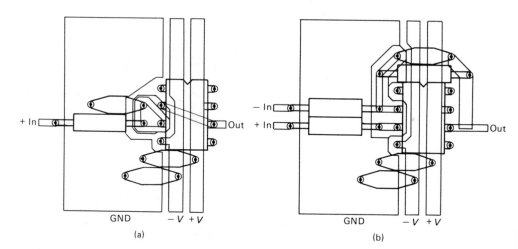

Figure 12-6 Printed circuit board layout for analog circuits: (a) noninverting buffer, (b) inverting amplifier

excessive ground inductance. Ringing on the falling edges of signals indicates excessive power/ground inductance. Also perform some preliminary emissions tests. By spending one day checking the current board design, you should be able to correct all these power/ground problems in the redesign.

In October 1983 I started designing the tester for a new printer board. After three months of hard work, the tester was qualified and ready for production. Then, in January 1984, the designer of the printer board told me that the product had failed the FCC emissions tests, and they would have to re-layout the board. The new layout took two months and forced me to redesign my tester. In May I had the new tester working, when guess what! The new layout also failed the emissions tests! This time the EMC lab came up with a fix which required adding six inductors to the board. To meet our start-of-production date we had to modify several thousand boards, cutting lands and adding the inductors by hand. This board went through two more layouts (without affecting my tester, fortunately) before we had a completely satisfactory design. If we had been a little smarter in our initial design, or spent more time characterizing the first design before we started the re-layout, we might have saved thousands of dollars and several months of agony for all concerned.

RECOMMENDED READING

BLOOD, WILLIAM R., JR., *MECL System Design Handbook*, 4th ed. Phoenix, AZ: Motorola Semiconductor Products Inc., 1983.

KEENAN, R. KENNETH, *Decoupling and Layout of Digital Printed Circuits*. Pinellas Park, FL: TKC, 1985.

KEENAN, R. KENNETH, *Digital Design for Interference Specifications*. Pinellas Park, FL: TKC, 1983.

MARDIGUIAN, MICHEL, *Interference Control in Computers and Microprocessor-Based Equipment*. Gainesville, VA: Don White Consultants, Inc., 1984.

VIOLETTE, MICHAEL F., and J. L. N. VIOLETTE, "EMI Control in the Design and Layout of Printed Circuit Boards." *EMC Technology*, 5:2 (March–April 1986), 19–20*ff*.

WHITE, DONALD R. J., *EMI Control in the Design of Printed Circuit Boards and Backplanes*. Gainesville, VA: Don White Consultants, Inc., 1981.

13

CABLES AND BACKPLANES

Cables and backplanes carry power, ground, and data between components in a subassembly, between subassemblies in a system, and between systems. These signals may range from microvolts to thousands of volts, from microamps to hundreds of amps, and from DC to several gigahertz, but the wiring is supposed to transport all the signals without significantly affecting them. To do this, the wiring must be properly designed, assembled, and installed.

We start by segregating signals according to their voltage, current, and frequency. To minimize crosstalk within a cable, the weakest signal should have at least one-fourth the voltage and one-fourth the current of the strongest signal. This usually separates the cables into six groups:

1. AC power, AC returns, chassis grounds, interfering audio signals and their returns;
2. DC power, DC returns, DC references, susceptible audio signals and their returns;
3. digital signals and their returns;
4. interfering RF signals and their returns;
5. susceptible RF signals and their returns; and
6. antenna signals.

Sensitive electronic systems need clean AC power. Regular branch circuits will usually work if they connect directly to main service panels and serve *only* the sensitive systems.

Miswired outlets can create serious problems, as shown by Figure 13-1. AC current flows through the chassis grounds and the interconnecting cable, because neutral and ground have been swapped between the two wall outlets. This high AC current is almost guaranteed to cause noise problems and could burn up the signal cable. (*Note:* Ecos Electronics sells ''power analyzers'' [$90 and up] that can identify every type of AC wiring error, while Radio Shack's ''outlet analyzer'' [$7] can identify open connections and swaps involving the hot wire.)

Chassis grounds and signals that create audio interference may be included in AC power cables. Always twist the AC power wires and the AC power returns together.

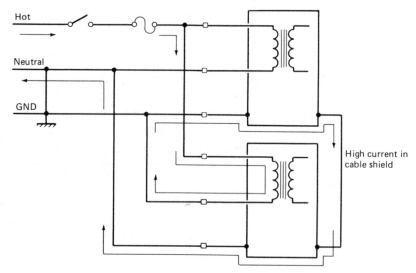

Figure 13-1   Incorrect power wiring that produces large ground loops

    Signals that are susceptible to audio interference may be bundled with DC power wires and DC power returns but must have their own signal returns. *Do not use DC power returns as chassis grounds or signal returns.* Twist DC power wires and their returns together. For multistage amplifiers, bring power in at the output stage. Shield the power cables going to noisy circuits, bonding the shields to the chassis at the ends of the cables and every 0.2λ in-between.

    The weakest signal in a cable should have at least one-fourth the voltage and one-fourth the current of the strongest signal in that cable. If you can, use separate cables for analog signals, digital signals, RF signals, antenna signals, and high-power signals. If you must mix signals within a cable, provide a signal return for each signal wire and use extra ground wires to help isolate the noisy signals from the sensitive signals.

    Each signal return should serve no more than nine signal wires, and one signal return per signal wire works much better. The best signal-ground arrangement for a two-layer backplane or signal cable is

> signal GND signal GND signal . . . GND signal
> GND signal GND signal GND . . . signal GND.

The best signal-ground arrangement for a single-layer cable is

> signal GND signal GND signal . . . GND signal.

Terminate each signal return at the signal source and at the load. Provide ground wires in the signal cables for switches and controls—do not connect them to chassis ground! If a signal cable has spare wires, tie half of them to ground at one end and the other half to ground at the other end. Run clock signals only where they are needed, and keep the clock wires and their returns very close together; a 0.001 m^2 clock loop can easily exceed the FCC limits for emitted radiation.

Keep isolated signal wires less than 0.15 meter long, and single wires near groundplanes less than 0.5 meter long. Single wires usually have ≈0.8 μH/m inductance (Appendix D) and 100–200 Ω impedance. Use wire with thick, low-permittivity insulation (Appendix C) to reduce crosstalk. Minimize the loop area (area enclosed by the source, signal wire, load, and return wire) in connectors, near other cables, and near electromagnetic devices. Flat cables and ribbon transmission-line cables are good up to 150 MHz, if they are kept away from metal and magnetic fields. Groundplane cables provide more shielding and better impedance control than cable without groundplanes. If possible, pair each signal with its signal return.

When signals are paired with signal returns, twisted-pair cables have low inductive coupling and constant impedance up to ≈10 MHz. Above 10 MHz the impedance may vary because of cable length, twist rate, and twist uniformity. The twist chord (length of one-half twist) should be less than one-tenth the distance to noisy/sensitive cables/components, and less than λ/4, where λ is the wavelength of the highest-frequency signal on the cable. Cables with over 13 twists per meter (most have 23–26 twists/meter) avoid standing-wave problems from DC to 1 GHz. Some twisted-pair cables have adjacent pairs twisted in opposite directions to reduce crosstalk. In most cases the largest loop area will be at the connectors. Connectors with conductive, grounded shells can greatly reduce radiation and noise pickup from these loops.

Inter-8 cable (trademark of Magnetic Shield Division of Perfection Mica) has four wires woven together to form very small, very uniform loops. This cable reduces radiation by a factor of 10, and noise pickup by a factor of 4, over equivalent twisted-pair cables.

Coaxial cables provide good noise control up to 100 MHz. To reduce high-voltage noise at low frequencies, ground the shield at the driver. To reduce high-voltage noise at high frequencies, ground the shield at the driver and every 0.2λ along the cable. Otherwise use the shield for the signal return. Above ≈50 kHz the shield current equals the signal current, minimizing inductive coupling. Above 1 MHz, coaxial cable acts like triaxial cable: signal-return currents flow on the inside of the shield, noise currents on the outside. Avoid coaxial cables with spiral-wrapped shields and cables whose shields have less than 95% optical coverage. Ground the shield with 360-degree bonds if you can. You may use "pigtails" below 1 MHz, but try to keep them less than 6 mm long. Keep exposed center conductors as short as possible ($\leq$ 13 mm). Drivers and receivers for coaxial cable should be designed for a nominal 100 Ω impedance. Triaxial cable should use the inner shield as the signal return and have the outer shield grounded at the driver, at the receiver, and every 0.2λ along its length.

Shielded twisted-pair cables provide superb protection against electric and magnetic fields up to 100 kHz and good protection up to 10 MHz. Signal and signal-return currents flow in the inner conductors, while noise currents flow only in the shield. Use double-shielded twisted-pair cables to carry audio signals in high-RF fields. Ground the outside shield at the ends and every 0.2λ in-between, and ground the inner shield at the signal source.

Most cables generate some electrostatic noise when they flex. If this is a problem, either clamp the cables in place, or use cables with a semiconductive layer deposited on the insulator. High-permeability tapes and coatings will limit RF propagation (25 kHz to 50 MHz) along cables.

Try to keep unfiltered AC power out of system enclosures. In Figure 13-2(a) the line filter's case is bonded to the chassis to form a continuous shield. In Figure 13-2(b) the AC power cord enters the chassis through a small hole and goes straight to the line filter. In either case, the line filter should have less than 1 mΩ resistance to chassis ground. If you are not using a line filter, the AC power cord should run straight to the load

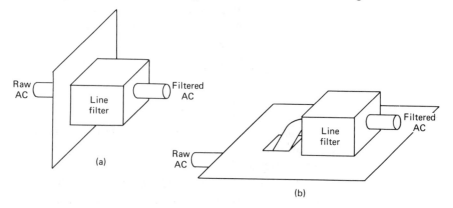

Figure 13-2 Line-filter installation

from a nearby hole in the chassis. Keep all other cables well away from this area, and keep AC power cables away from all other apertures in the chassis. If an AC power cord is longer than needed, bunch it up or pile it in a figure-8 shape.

Mount DC power cables close to chassis members and away from apertures, AC power cables, transformers, motors, and solenoids.

Mount signal cables close to chassis members and away from apertures, AC power cables, DC power cables, transformers, motors, and solenoids. Keep signal cables at least 0.15 m away from power cables. Separate digital-signal cables and AC power cables by at least one-fortieth of their parallel run in order to provide 1000–10,000:1 isolation from DC to 10 MHz. Separate analog-signal cables and AC power cables by at least one-fourth of their parallel run. Put 4 mm-thick foam spacers between long, parallel signal cables. Input/output cables to slow devices do not need to be shielded if the drivers and receivers are properly decoupled.

Shield all exterior signal cables going to high-speed devices. Use 360-degree bonds between the shields and the chassis/connector covers to achieve ≈0.5 mΩ bond resistance. Sensitive signals may need individual shields. Insulate these shields to prevent accidental grounding, and run the signals and their shields through adjacent connector pins. Ground the shield at only one point if shield currents would affect the signal: ground the source end to reduce noise emissions, and ground the load end to reduce noise pickup. Shields for RF signals should be grounded at the ends and at least every 0.2λ in-between.

Figure 13-3 shows the relative noise immunity of various cabling schemes when the signal source is grounded. Figure 13-4 shows the relative noise immunity of various cabling schemes when the source is floating. Figure 13-3(a) shows the reference circuit: a single signal wire. In Figures 13-3(c) through (f) we reduce the noise emissions and noise pickup by reducing the effective area of the current loop. Floating the source or the load (Figure 13-4(a) through (e)) forces all of the signal current to return through the cable shield/return conductor, greatly reducing the effective area and the noise. Notice, however, that just twisting a signal-return line around the signal wire (Figure 13-3(d)) can reduce noise emissions and noise pickup by 75%.

An easy way to shield a cable is to wrap it with 50 to 100 mm-wide strips of aluminum foil, spiral-wrapped around the cable and secured with tape or tie wraps. Quite a few companies manufacture copper tape, aluminum tape, knitted-wire tapes, and high-permeability tapes for this

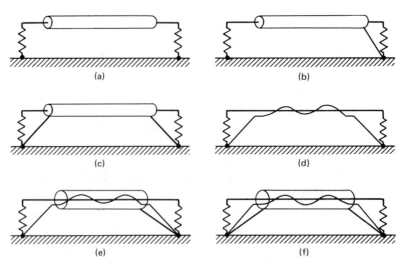

Figure 13-3   Noise immunity of cables with grounded signal sources: (a) 1:1 (ref),
(b) 1:1, (c) 2–22:1, (d) 4:1, (e) 1–4:1, (f) 25:1

same purpose. A few companies produce shielding tubes and accessories
that can be zipped onto existing cables.

In autumn of 1983, one of my co-workers came to me with a
problem. He was designing a functional tester for a new typewriter and
had noise problems with some analog measurements. He was trying to
measure the voltage across some current-sense resistors (300 mV with

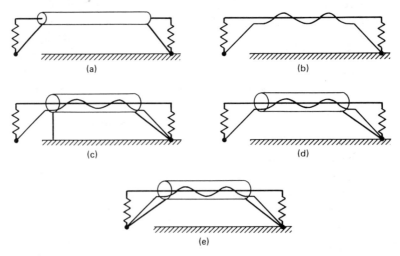

Figure 13-4   Noise immunity of cables with floating signal sources: (a) 280–10,000:1,
(b) 4–8900:1, (c) 1500:1, (d) 1600–3200:1, (e) 3500–7100:1

75 mV ripple) and was getting completely erroneous results. When we looked at the signals with an oscilloscope we saw 2 V spikes on the current-sense signals. Tracing through the wiring, I discovered that the motor-drive signals (38 V at 1 A, pulse-width-modulated at 20 kHz) and the current-sense signals ran through the same cable. I suggested splitting this cable into two twisted-pair cables, one for the motor-drive signals and one for the current-sense signals, entering the tester through separate connectors. We made this change and arranged the cabling inside the tester to provide 100 mm spacing between these signals. This time, when we looked at the current-sense signals, we didn't see any spikes at all, and the measurements were within a few percent of the expected values.

RECOMMENDED READING

Bell Telephone Laboratories, *Physical Design of Electronic Systems*, Vol. 1. Englewood Cliffs, NJ: Prentice-Hall, Inc., 1970.

DEMOULIN, BERNARD, and PIERRE DEGAUQUE, "Effect of Cable Grounding on Shielding Performance," *EMC Technology*, 3:4 (October–December 1984), 65–69*ff*.

KEISER, BERNHARD E., *EMI Control in Aerospace Systems*. Gainesville, VA: Don White Consultants, Inc., 1979.

MARDIGUIAN, MICHEL, *How to Control Electrical Noise*. Gainesville, VA: Don White Consultants, Inc., 1983.

WHITE, DONALD R. J., *EMI Control in the Design of Printed Circuit Boards and Backplanes*. Gainesville, VA: Don White Consultants, Inc., 1981.

14

SHIELDING

All too often, expensive shields are tacked onto systems to cure noise problems that should have been avoided in the system design. For example, if a regular 60-Hz power transformer is mounted near the cathode-ray tube (CRT) in an oscilloscope, the trace will jitter up and down. One way to solve this problem is to add a costly three-layer shield. Or, if we had considered this problem from the beginning, we could have (1) chosen a low-leakage transformer; (2) mounted the transformer on the underside of a steel chassis, with a steel cover plate to form a shielded enclosure; (3) mounted the transformer well away from the CRT, with its leakage field aligned with the long axis of the CRT; and (4) left enough room around the CRT for a standard single-layer shield if needed.

Shielding effectiveness (attenuation) is usually specified in decibels (dB), defined by

$$SE = 10 \log_{10} \left(\frac{\text{power without shield}}{\text{power with shield}} \right) \text{dB}.$$

For a shield in air or vacuum, this becomes

$$SE = 20 \log_{10} \left(\frac{E\text{-field without shield}}{E\text{-field with shield}} \right) \text{dB}$$

$$= 20 \log_{10} \left(\frac{H\text{-field without shield}}{H\text{-field with shield}} \right) \text{dB}.$$

Most noise and interference problems require 30–60 dB shielding. Above 100 kHz, single-layer shields can provide 40–70 dB shielding and double-layer shields can provide up to 120 dB shielding. Holes and gaps usually limit the attenuation of electric fields and high-frequency magnetic fields, while the shield material and shield thickness limit the attenuation of low-frequency magnetic fields. (*Note:* To attenuate magnetic fields, the shield must remain unsaturated.)

Figure 14-1 shows an electromagnetic wave with frequency f (Hz) and impedance $| Z_w |$, hitting a shield of thickness T (m), permeability $\mu = \mu_r \mu_v \approx 1.257 \mu_r$ μH/m, and resistivity ρ (Ω-m). At this frequency the shield has impedance

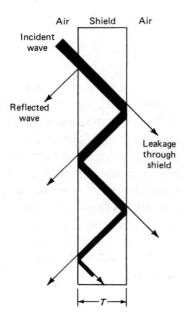

Air   Shield   Air

Incident
wave

Reflected
wave

Leakage
through
shield

$\leftarrow T \rightarrow$

Figure 14-1 Reflections and dissipation within the shield attenuate of incident fields

$$|Z_s| = \frac{(2\,\pi\mu f\rho)^{1/2}}{1 - \exp\left(-\dfrac{T}{\delta}\right)} \;\Omega/\text{square},$$

where

$$\delta = \left(\frac{\rho}{\pi\mu f}\right)^{1/2} \text{meters}.$$

The impedance mismatch between the wave and the shield causes part of the wave to reflect off the air-shield interface. The rest of the wave enters the shield, where some of it gets turned into heat. When it reaches the shield-air interface, part of the wave leaks out and the remainder gets reflected back into the shield. This re-reflected wave continues to bounce back and forth inside the shield until it leaks out or is turned into heat. We can express the shielding effectiveness of a solid sheet as

$$SE = R + A + B \text{ dB},$$

where

$$R = -20\log_{10}\left(\frac{4\,|Z_s|\,|Z_w|}{(|Z_s| + |Z_w|)^2}\right) \text{dB},$$

$$A = -20 \log_{10} \left(\exp \left[-\frac{T}{\delta} \right] \right) \text{ dB} \approx 8.69 \frac{T}{\delta} \text{ dB},$$

$$B = +20 \log_{10} \left(1 - \left(\frac{|Z_s| - |Z_w|}{|Z_s| + |Z_w|} \right)^2 \exp \left[-2\frac{T}{\delta} \right] \right) \text{ dB}.$$

R is the reflection term, A is the attenuation term, and B is the re-reflection term. If $A \geq 10$ dB, $B \approx 0$ dB. Table 14-1 shows the skindepth (δ) for common shielding materials over the range 50 Hz to 10 GHz, while Table 14-2 compares the shield impedance $|Z_s|$ of thin (25.4 μm) to thick (1 mm) shields over the same range. Notice that when $T > 3\delta$, the shield impedance is independent of the shield thickness. In general, $|Z_s|$ should be less than 1 Ω/square for a shield to be effective ($\rho < 0.05$ Ω-m).

The high impedance of electric fields and plane waves ($|Z_w|$ ≥ 376.7 Ω, see Appendix H) makes them reflect off the air-shield interface, so shields can be very thin and still be effective. By grounding the shield we can also block low-frequency electric fields. Tables 14-3 and 14-4 show the shielding effectiveness of various materials against 4 kΩ electric fields and plane waves.

The low impedance of magnetic fields ($|Z_w|$ < 376.7 Ω, see Appendix H) makes them penetrate the air-shield interface and reflect off the shield-air interface. Through multiple internal reflections any portion of the field that is not absorbed by the shield will eventually leak out. Thus shields for magnetic fields must be thick with respect to skindepth δ, with high permeability μ and low resistivity ρ (see Appendix B). Table 14-5 shows the shielding effectiveness of various materials against 40

TABLE 14-1 Skindepth of Common Shielding Materials

Frequency	Aluminum	Copper	Mu-Metal	Steel	Zinc
50 Hz	11.7 mm	9.33 mm	248 μm	1.20 mm	17.4 mm
100 Hz	8.27 mm	6.60 mm	175 μm	851 μm	12.3 mm
1 kHz	2.62 mm	2.09 mm	55.4 μm	269 μm	3.90 mm
10 kHz	827 μm	660 μm	17.5 μm	85.1 μm	1.23 mm
100 kHz	262 μm	209 μm	5.54 μm	26.9 μm	390 μm
1 MHz	82.7 μm	66.0 μm	1.75 μm	8.51 μm	123 μm
10 MHz	26.2 μm	20.9 μm	554 nm	2.69 μm	39.0 μm
100 MHz	8.27 μm	6.60 μm	175 nm	851 nm	12.3 μm
1 GHz	2.62 μm	2.09 μm	55.4 nm	269 nm	3.90 μm
10 GHz	827 nm	660 nm	17.5 nm	85.1 nm	1.23 μm

TABLE 14-2 Impedance of 25.4 μm- to 1 mm-Thick Shields in Ohms/Square

Frequency	Aluminum	Copper	Mu-Metal	Steel	Zinc
50 Hz	1.5 mΩ–40 μΩ	960 μΩ–26 μΩ	34 mΩ–3.3 mΩ	7.9 mΩ–290 μΩ	3.3 mΩ–87 μΩ
100 Hz	1.5 mΩ–41 μΩ	960 μΩ–26 μΩ	34 mΩ–4.7 mΩ	7.9 mΩ–340 μΩ	3.3 mΩ–88 μΩ
1 kHz	1.5 mΩ–46 μΩ	960 μΩ–31 μΩ	40 mΩ–15 mΩ	8.2 mΩ–750 μΩ	3.4 mΩ–96 μΩ
10 kHz	1.5 mΩ–66 μΩ	980 μΩ–47 μΩ	61 mΩ–46 mΩ	9.0 mΩ–2.3 mΩ	3.4 mΩ–124 μΩ
100 kHz	1.6 mΩ–150 μΩ	1.0 mΩ–120 μΩ	150 mΩ	12 mΩ–7.4 mΩ	3.5 mΩ–240 μΩ
1 MHz	1.8 mΩ–460 μΩ	1.2 mΩ–370 μΩ	460 mΩ	25 mΩ–23 mΩ	3.7 mΩ–690 μΩ
10 MHz	2.4 mΩ–1.5 mΩ	1.7 mΩ–1.2 mΩ	1.5 Ω	74 mΩ	4.6 mΩ–2.2 mΩ
100 MHz	4.8 mΩ–4.6 mΩ	3.8 mΩ–3.7 mΩ	4.6 Ω	230 mΩ	7.9 mΩ–6.9 mΩ
1 GHz	15 mΩ	12 mΩ	15 Ω	740 mΩ	22 mΩ
10 GHz	45 mΩ	37 mΩ	46 Ω	2.3 Ω	69 mΩ

TABLE 14-3   Shielding Effectiveness of 25.4 μm- to 1 mm-Thick Shields
against Electric Fields ($Z_w \approx 4$ kΩ)

Frequency	Aluminum	Copper	Mu-Metal	Steel	Zinc
50 Hz	69–133	75–138	76–145	75–136	59–122
100 Hz	72–136	78–141	79–156	78–139	62–125
1 kHz	82–145	88–150	88–≥180	87–155	72–135
10 kHz	92–153	98–159	96–≥180	97–≥180	82–143
100 kHz	102–170	108–180	116–≥180	105–≥180	92–155
1 MHz	111–≥180	117–≥180	≥180	118–≥180	101–≥180
10 MHz	120–≥180	125–≥180	≥180	165–≥180	110–≥180
100 MHz	133–≥180	142–≥180	≥180	≥180	120–≥180
1GHz	≥180	≥180	≥180	≥180	150–≥180
10 GHz	≥180	≥180	≥180	≥180	≥180

TABLE 14-4   Shielding Effectiveness of 25.4 μm- to 1 mm-Thick Shields
against Plane-Waves ($Z_w = 376.7$ Ω)

Frequency	Aluminum	Copper	Mu-Metal	Steel	Zinc
50 Hz	49–112	55–118	55–124	54–116	38–102
100 Hz	52–115	58–121	58–136	57–118	41–105
1 kHz	62–124	68–130	67–≥180	67–134	51–114
10 kHz	72–133	77–139	76–≥180	76–≥180	61–123
100 kHz	81–149	87–160	96–≥180	85–≥180	71–134
1 MHz	91–≥180	96–≥180	172–≥180	98–≥180	80–173
10 MHz	99–≥180	105–≥180	≥180	144–≥180	89–≥180
100 MHz	112–≥180	121–≥180	≥180	≥180	99–≥180
1 GHz	161–≥180	≥180	≥180	≥180	129–≥180
10 GHz	≥180	≥180	≥180	≥180	≥180

TABLE 14-5   Shielding Effectiveness of 25.4 μm- to 1 mm-Thick Shields
against Magnetic Fields ($Z_w \approx 40$ Ω)

Frequency	Aluminum	Copper	Mu-Metal	Steel	Zinc
50 Hz	30–93	35–98	36–105	35–96	20–82
100 Hz	32–96	38–101	39–116	38–99	22–85
1 kHz	42–105	48–110	48–≥180	47–115	32–95
10 kHz	52–113	58–119	56–≥180	57–175	42–103
100 kHz	62–130	68–140	76–≥180	65–≥180	52–115
1 MHz	71–≥180	77–≥180	153–≥180	78–≥180	61–154
10 MHz	80–≥180	85–≥180	≥180	125–≥180	70–≥180
100 MHz	93–≥180	102–≥180	≥180	≥180	80–≥180
1 GHz	141–≥180	164–≥180	≥180	≥180	110–≥180
10 GHz	≥180	≥180	≥180	≥180	≥180

Ω magnetic fields. To prevent saturating a magnetic shield, leave $\geqslant 6$ mm clearance between components and the shield. Shields formed from foil should have $\geqslant 13$ to 20 mm overlap at seams, and the inside radius of bends should be at least twice the shield thickness.

In general, a shield thick enough to support itself will provide ample shielding against everything except low-audio magnetic fields. As a rule of thumb, use copper and aluminum to shield against electric fields, plane waves, and magnetic fields above 1 MHz. Use steel or iron foil to shield against magnetic fields from 10 kHz to 1 MHz, and use high-permeability alloys (Mu-Metal, Permalloy) to shield against magnetic fields below 10 kHz when the shield size and weight are critical. (*Note:* High-permeability alloys are extremely sensitive to shock—a 0.6-m drop can cut the magnetic shielding effectiveness in half!) If we add a shield to reduce noise emissions from a circuit, we may encounter problems because of resonances inside the shield, especially when the longest shield dimension approaches half a wavelength. To avoid this problem, we can make the shield out of low-resistivity high-permeability metal in order to absorb the fields instead of reflecting them.

Aluminum and copper tapes are available for shielding inductors, transformers, and cables and for sealing seams in shields ($11 to $22 per m^2). The most effective shielding tape is embossed copper, followed by embossed aluminum, copper with conductive adhesive, aluminum with a conductive adhesive, smooth copper, and smooth aluminum. Ordinary aluminum foil (≈ 25 μm thick) is an effective shield against electric fields and $\geqslant 100$ kHz magnetic fields.

Shields for low-audio magnetic fields usually have a layer of copper, a layer of high-saturation low-permeability metal, and a layer of low-saturation high-permeability metal. Multilayer shields work best if they have small air gaps (0.5 to 0.75 mm or the metal thickness, whichever is greater) between the layers. All shields using high-permeability metals require special heat treatment and may lose their shielding properties if they are formed, machined, suddenly chilled, or dropped. These shields should be completely formed by the manufacturer and carefully handled thereafter.

Some companies sell shielding foils made of high-permeability metals. You can cut this foil with scissors (leaving wide overlaps for the joints), gently bend it to shape, and tape it together to form shields. When using this foil, make sure that the seams are aligned with the magnetic field. Keep the bends gentle, and do not weld or jar the finished shield. These foils can also be diecut to form gaskets and shields. Flexible shielding laminates combine foil with a reinforcing insulator. They cost $3 to $14/m^2 and can be diecut to shape. Flexible conductive fabrics are also available and can be used to shield cables, connectors, and CRTs or formed into grounding straps ($20 to $40/m^2).

A fairly common problem nowadays is how to shield molded plastic enclosures. One solution is to use conductive fillers in the plastic, providing 30 to 80 dB shielding at a cost of $6 to $11/m^2. The main drawbacks are the degradation in the mechanical properties of the plastic, and non-uniform dispersion of the filler in the plastic. Two other problems are making good contact, since the surfaces tend to be resin-rich, and trying to achieve uniform color—these enclosures usually must be painted in a separate operation.

Another solution is to apply a conductive coating to the plastic. A 4 to 5 μm layer of vacuum-metallized aluminum can be applied for $8 to $38/m^2. This entails a $100,000 to $150,000 investment in chambers; a base coat is required for good adhesion; and the coating may corrode in humid environments. Sputtering can deposit a ≈1.5 Ω/square coating for $11 to $22/m^2. This entails an investment of $300,000 to $3,000,000 in chambers. A 50 to 75 μm coating of copper, nickel, or silver paint may be sprayed on for $6–$32/m^2, providing 30–65 dB shielding. This is the easiest method to use for prototypes and low-volume products. Zinc-arc and zinc-flame spray can apply a 50 to 125 μm layer of zinc for $10 to $22/m^2. The zinc has ≈20 mΩ/square resistance and provides 70 to 90 dB shielding. The complete setup costs $10,000 to $20,000. The plastic must first be gritblasted, or sprayed with a tie-coat, to provide "tooth" and keep the zinc from flaking off with temperature changes. The fairly high application temperatures may warp the plastic, and surface treatment may be needed to protect the zinc from damage. Electroless plating can apply nickel over copper, providing 55 to 110 dB shielding, for $5 to $17/m^2.

Transparent conductive coating can be applied to CRTs and viewing windows, with tin oxide, indium tin oxide, and gold being the most common. These coatings should have less than 1 Ω/square resistivity and be designed with 6 dB safety factor to cover manufacturing variations. Another technique is to embed a fine metal mesh in the viewing window. In either case, gaskets will probably be needed to bond the conductive coating/metal mesh to the chassis.

Most of the leakage through shields is through gaps, holes, and nonhomogeneous areas. These discontinuities are effectively in parallel with the shield and must be designed and built to minimize leakage.

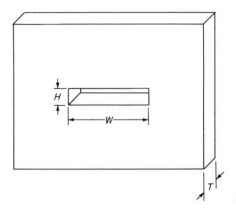

Figure 14-2   Rectangular slot in a shield

Figure 14-2 shows a rectangular slot in a shield, such as a cooling slot or a formed edge of the shield. If this slot is W meters wide by H meters high ($W \geqslant H$), and the shield is T meters thick, its shielding effectiveness against a signal with wavelength $\lambda > 2(W + H)$ meters is

$$
\begin{aligned}
SE = &-20 \log_{10} \left(\frac{(WH)^{1/2}}{0.24\lambda} \right) \\
&-20 \log_{10} \left[\exp \left(-\frac{\pi T}{W} \left[1 - \left(\frac{2w}{\lambda} \right)^2 \right]^{1/2} \right) \right] \text{ dB} \\
\approx &-20 \log_{10} \left[\frac{(WH)^{1/2}}{0.24\lambda} \right] + 27.3 \left(\frac{T}{W} \right) \text{ dB} \quad \text{for } \lambda \gg 2W.
\end{aligned}
$$

Figure 14-3 shows a round hole in a shield, such as control shaft, meter face, or fan opening. If this hole is D meters in diameter and the shield is T meters thick, its shielding effectiveness against a signal with wavelength $\lambda > \pi D$ meters is

$$
\begin{aligned}
SE = &-20 \log_{10} \left[\frac{D}{0.3\lambda} \right] \\
&-20 \log_{10} \left[\exp \left(-\frac{2\pi T}{1.707D} \left[1 - \left(\frac{1.707D}{\lambda} \right)^2 \right]^{1/2} \right) \right] \text{ dB} \\
\approx &-20 \log_{10} \left[\frac{D}{0.3\lambda} \right] + 32.0 \left(\frac{T}{D} \right) \text{ dB} \quad \text{for } \lambda \gg \pi D.
\end{aligned}
$$

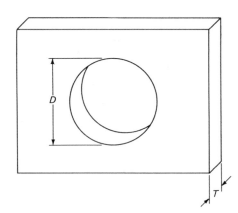

Figure 14-3   Round hole in a shield

The first term in these equations is the aperture effect and the second term is the waveguide-below-cutoff effect. In general, slots should be less than 0.05λ meters wide, aligned parallel to current in the shield, and holes should be less than 0.05λ meters in diameter. The shielding effect of groups of apertures also depends on the distance from the signal source to the shield and on the spacing between the holes.

Figures 14-4(a) and (b) show T-meter thick shields with arrays of rectangular slots, W meters by H meters ($W \geqslant H$) with walls S meters wide between the holes. The shields are made out of metal with resistivity ρ (Ω-m) and permeability $\mu = \mu_v \mu_r \approx 1.257 \mu_r$ μH/m. The signal source is r meters from the shield and produces a signal of frequency f hertz

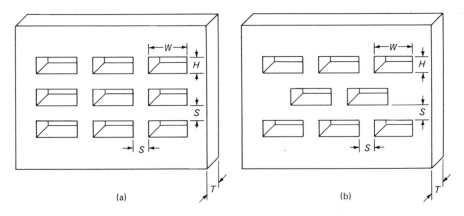

Figure 14-4   Array of rectangular slots in a shield

and wavelength $\lambda = c/f \approx 2.998 \times 10^8/f$ meters. The shielding effectiveness of the slots is

$$SE = A_a + R_a + B_a + K1 + K2 + K3 \text{ dB},$$

where

$$A_a = 27.3 \frac{T}{W} \text{ dB};$$

$$R_a = 20 \log_{10} \left(\frac{|1 + k|^2}{4|k|} \right) \text{ dB, and}$$

$$B_a = 20 \log_{10} \left| 1 - \frac{(k - 1)^2}{(k + 1)^2} 10^{-2.73T/W} \right| \text{ dB},$$

$$k = \frac{W}{\pi r} \text{ for magnetic fields,}$$

$$k = \frac{j2W}{\lambda} \text{ for plane waves,}$$

$$k = -\frac{4\pi Wr}{\lambda^2} \text{ for electric fields;}$$

$$K1 = 10 \log_{10} \left(\frac{(W + S)(H + S)}{WH} \right) \text{ dB};$$

$$K2 = -20 \log_{10} \left[1 + \frac{35}{\left(\frac{S}{\delta} \right)^{2.3}} \right] \text{ dB},$$

where

$$\delta = \left(\frac{\rho}{\pi \mu f} \right)^{1/2} \text{ meters};$$

$$K3 = 20 \log_{10} \left[\frac{\exp(6.29T/W) + 1}{\exp(6.29T/W) - 1} \right] \text{ dB}.$$

If the signal source is far away from the shield ($r \gg W + S$ meters) use the $K1$ factor above. As the signal source approaches the shield, the exact position of the source and the closest hole becomes important. To be safe, use $K1 = 0$ dB when the noise source/receiver is close to the shield.

Figures 14-5(a) and (b) show similar shields with arrays of round holes of diameter D meters. The shielding effectiveness of the holes is

$$SE = A_a + R_a + B_a + K1 + K2 + K3 \text{ dB},$$

where

$$A_a = 32.0 \frac{T}{D} \text{ dB};$$

$$R_a = 20 \log_{10} \left(\frac{|1 + k|^2}{4|k|} \right) \text{ dB, and}$$

$$B_a = 20 \log_{10} \left| 1 - \frac{(k - 1)^2}{(k + 1)^2} 10^{-3.20T/D} \right| \text{ dB},$$

where

$$k = \frac{D}{3.682r} \text{ for magnetic fields,}$$

$$k = \frac{j2\pi D}{3.682\lambda} \text{ for plane waves,}$$

$$k = -\frac{4\pi^2 Dr}{3.682\lambda^2} \text{ for electric fields;}$$

$$K1 = 10 \log_{10} \left(\frac{4(D + S)^2}{\pi D^2} \right) \text{ dB for straight lines of holes, (Figure 14-5(a)), or}$$

$$K1 = 10 \log_{10} \left(\frac{3.464(D + S)^2}{\pi D^2} \right) \text{ dB for staggered lines of holes (Figure 14-5(b));}$$

$$K2 = -20 \log_{10} \left[1 + \frac{35}{\left(\frac{S}{\delta} \right)^{2.3}} \right] \text{ dB},$$

where

$$\delta = \left(\frac{\rho}{\pi \mu f} \right)^{1/2} \text{ meters;}$$

$$K3 = 20 \log_{10} \left[\frac{\exp{(7.37T/D)} + 1}{\exp{(7.37T/D)} - 1} \right] \text{dB}.$$

Again, if the noise source is close to the shield, use $K1 = 0$ dB to be safe.

For wire mesh (Figure 14-6), use the equations for a shield with an array of rectangular slots, letting $T = S =$ the diameter of the wire, and letting $K1 = 0$ dB. For metal honeycomb, use the equations for a shield with round holes, letting D be the distance across flats, and letting S be the wall thickness.

From the discussion above, you can see the importance of keeping seams as tight as possible. Continuous-weld seams (Figure 14-7(a), (b), (c)) work best; the welding rod should match the base metal (especially the carbon content) to provide a homogeneous bond. Seams that are overlapped $\geqslant 10$ mm and spot-welded at least every 50 mm (preferably $\leqslant 13$ mm, Figure 14-7(d)) are next best. A rolled-and-overlapped seam (Figure 14-7(e)) also works well and can be crimped at 50 mm intervals

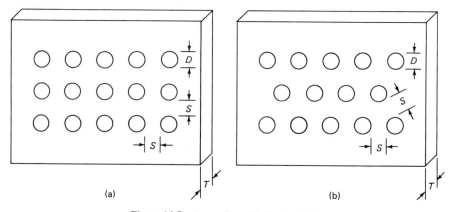

Figure 14-5   Array of round holes in a shield

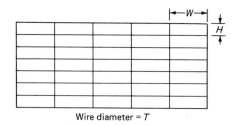

Wire diameter = T

Figure 14-6   Wire mesh

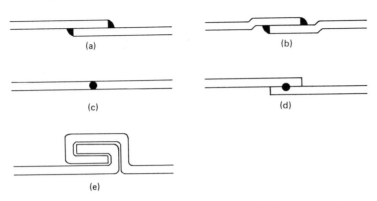

Figure 14-7 Permanent seams in shields: (a) continuous weld, (b) stepped continuous weld, (c) butt weld, (d) spot welds, (e) overlapped seam

to ensure solid contact. When forming these seams, overlap the edges as much as possible. Openings in permanent seams can be filled with conductive paste or conductive caulk to reduce leakage. Riveted seams should have rivets at least every 20 mm, and the rivet holes should be a close fit to the rivets. If possible, arrange the seams in a shield to parallel the current flow in the shield.

Covers and removable panels require seams that can be opened for maintenance. Flat cover plates (Figure 14-8(a), (b), and (d)) tend to buckle slightly, forming slots with H/W ratios near 0.002. Forming lips on covers (Figure 14-8(c)) makes them stiffer, reducing the H/W ratio to ≈ 0.0002. (*Note:* To reduce buckling when fastening covers, start at the center of each edge and work toward the corners.) The fasteners should be $\leq 0.05\lambda$ meters apart for commercial equipment and $\leq 0.02\lambda$ meters apart for military equipment. The overlap between the cover and the main shield

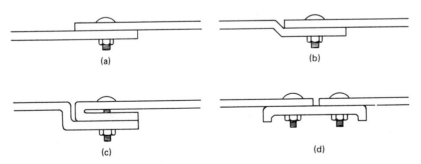

Figure 14-8 Temporary seams in shields: (a) overlapped, (b) stepped, (c) stepped and folded, (d) channel

should be ⩾2.5 times the cover thickness and ⩾5 times the maximum gap in order to enhance the waveguide-below-cutoff effect. Keep noisy/sensitive circuits away from the edges of covers, and consider using grounding straps, fingerstock, or EMI gaskets to help bond the covers to the chassis.

Gaskets help seal poorly fitting joints (Figure 14-9) but may have eight times the impedance of direct metal-to-metal bonds. Gaskets should be mounted inside the clamping bolts (Figure 14-9(g) and (h)) to prevent leakage through the bolt holes (Figure 14-9(a) and (b)). The joint and gasket design should provide 34–138 kPa pressure on every section of the gasket while avoiding overcompression and should keep the members from sliding across one another. When an EMI gasket is combined with an environmental seal, the seal should face the harsh environment. Gaskets can be held in place by welded or riveted strips (Figure 14-9 (c), (d), (e), (f)) or they may be glued in place with conductive adhesives (Figure 14-9(a), (b), (g), (h)). The conductive adhesive should have ⩽100 $\mu\Omega$-m resistivity and be applied as single 3–6 mm dots every 25–50 mm.

The gasket material should be chosen for resilience, corrosion resistance (more cathodic than the main structure), and low resistivity. Monel, silver-plated brass, and aluminum woven-wire gaskets can provide 54 dB attenuation up to 1 GHz, when clamped with 138 kPa pressure.

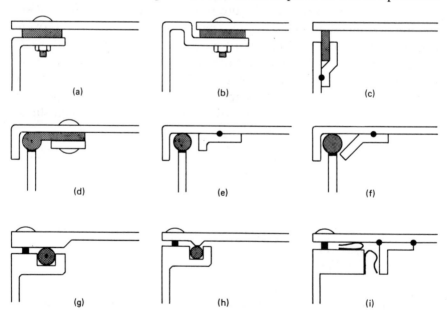

Figure 14-9   Sealing temporary seams with gaskets

Wire sleeving over a neoprene rod is also very good and is less sensitive to the clamping pressure. Phosphor-bronze or beryllium-copper fingerstock can provide 60 dB attenuation from 10 kHz to 10 GHz if the fingers are spaced less than $\lambda/4$ meters apart and clamped with 5–10 grams pressure each. Multiple rows of fingerstock are necessary to shield against magnetic fields; half of the fingerstock should be compressed by the joint, and the other half wiped (Figure 14-9(i)).

The gasket height depends on how many times the joint will be assembled and the joint unevenness (maximum gap with the members just touching). A gasket that will be installed once and then forgotten should be twice the joint unevenness. A door or access panel, that keeps the members in the same relative positions, should have a gasket three times as thick as the joint unevenness. Gaskets that may be removed and re-installed should be four times as thick as the joint unevenness.

One problem with high-frequency equipment is how to install controls and indicators without losing shielding effectiveness. Toggle switches usually present no problems, but panel switches may need direct metal-to-metal bonds to the panel. Rotary switches may use a conductive packing in the bearing, brushes on the shaft next to the panel, a fingerstock ring around the shaft, or a metal-mesh gasket in a compression gland. CRTs and meters may need to be installed inside shields bonded to the back of the panel, with the signals being brought through feedthrough capacitors. Shielding windows using metallized fabric, wire mesh, or transparent conductive coatings may also be used, but they tend to be expensive and hard to see through.

Another method is to use the waveguide-below-cutoff effect. A rectangular tube three times as long as it is wide ($T \geq 3W \geq 3H$) attenuates signals with wavelength $\lambda > 2W$ by 82 dB. A round tube whose length is three times its diameter ($T \geq 3D$) attenuates signals with wavelength $\lambda > 1.707D$ by 96 dB. (The shortest wavelength used in the system should be at least double these critical wavelengths.) To form a maintenance-free seal for a rotary switch, just weld or braze a long tube to the panel and run a nonconductive shaft through the tube. This same idea can be applied to high-frequency amplifiers—design the chassis and shield to form a long, narrow waveguide (W meters wide, H meters high, $W \geq H$). If a stage has x dB gain, separate it from the preceding and following stages by $\geq xW/27.3$ meters, and you will have *no* unwanted inductive coupling between the stages. Similarly, a tubular CRT shield can be bonded to the front panel to avoid the need for a shielding window.

Shields should have separate openings for incoming signals and

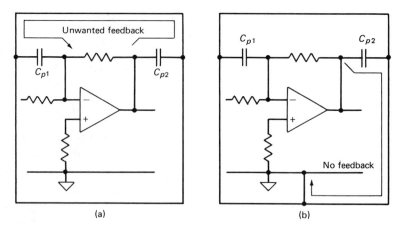

Figure 14-10 Why a shield must connect to the zero-reference point of the circuitry inside: (a) floating shield, (b) shield connected to zero-reference

signal returns, for outgoing signals and their signal returns, and for power lines and power returns. Ventilation openings should be covered with expanded-metal grids, metal honeycomb, or metal mesh, bonded to the shield every 100 mm. Connections to the inside of a shield and to the outside of a shield should be separated by ≥ 10 times the shield thickness and ≥ 5 times the bond diameter. Shield grounds should never penetrate the shield and should be close to the signal entry/exit points in order to minimize noise currents in the shield.

A shield should connect to the zero-reference point of the circuitry inside it. If the shield floats, parasitic capacitance (C_{p1}, C_{p2} in Figure 14-10(a)) can couple the circuit's input to the circuit's output, causing oscillations or other problems. By grounding the shield (Figure 14-10(b)), any current in the shield gets drained to ground, effectively breaking the feedback path. A sensitive system may need one shield per zero-reference (Figure 14-11(a)) and one shield per power entrance point. If a system has high common-mode voltages, you may need to add safety shields (Figure 14-11(b)) to protect operators and maintenance personnel.

In early 1984 a co-worker asked me to look at a noise problem in a linecord tester. About one time out of a hundred, the tester would reset itself in the middle of the test. Looking at the reset signal, we saw ≈ 4 V spikes at the beginning of the hi-pot (insulation-resistance) test. The reset wire ran from the microprocessor, across the top of the tester (parallel to the linecord being tested), to a switch connected to chassis ground. When the tester applied 3500 V to the linecord, some of the signal coupled

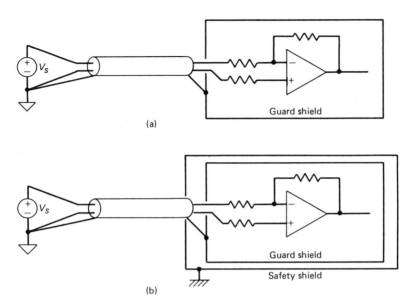

Figure 14-11   Grounding guard shields and safety shields

into the reset line through the parasitic capacitance. We chopped the noise spikes to ≈ 0.2 V and eliminated the noise problem by (1) running a ground wire to the switch in the signal cable, to help cancel the induced noise, and (2) covering the signal cable with $\frac{3}{8}$-inch tubular copper braid, grounded to the chassis near the microprocessor, to form an electrostatic shield.

RECOMMENDED READING

EMI Shielding Engineering Handbook. Woburn, MA: Chomerics, Inc., 1985.

Interference Reduction Guide for Design Engineers, Vol. 1. Springfield, VA: NTIS (AD 619 666), 1964.

COWDELL, ROBERT B., "Simplified Shielding," *1967 IEEE Electromagnetic Compatibility Symposium Record*, Washington, D.C., July 18–20, 1967, 399–412.

COWDELL, ROBERT B., "Simplified Shielding for Perforated Shields," *1968 IEEE Electromagnetic Compatibility Symposium Record*, Seattle, WA, July 23–25, 1968, 308–316.

MORRISON, RALPH, *Grounding and Shielding Techniques in Instrumentation*. New York: John Wiley & Sons, Inc., 1967.

OTT, HENRY W., *Noise Reduction Techniques in Electronic Systems*. New York: John Wiley & Sons, Inc., 1976.

SCHELKUNOFF, S. A., *Electromagnetic Waves*. Princeton, NJ: D. Van Nostrand Co., Inc., 1943.

WHITE, DONALD R. J., *A Handbook on Electromagnetic Shielding Materials and Performance*, 2nd ed. Gainesville, VA: Don White Consultants, Inc., 1980.

WHITE, DONALD R. J., *A Handbook Series on Electromagnetic Interference and Compatibility*, Vol. 3. Gainesville, VA: Don White Consultants, 1973.

WHITE, DONALD R. J., *Shielding Design Methodology and Procedures*. Gainesville, VA: Interference Control Technologies, 1986.

15

FILTERING

Just like shields, filters are frequently added to systems to cure noise problems that should have been avoided in the first place. These last-minute "band-aids" can be very expensive, frequently adding 10%–15% to a system's cost, whereas the same amount of noise protection could have been designed in for ≈1% of the base system cost. This 9%–14% cost hit is mainly due to the greater noise attenuation required by poorly designed systems, the need to cram the additional filters into restricted spaces, and grounding problems that reduce their effectiveness. For example, a 37-pin filtered connector costs about $37 in quantities of 100, while a similar unfiltered connector costs about $4.

By considering filtering as part of the system design we can provide solid grounds and good input-output isolation, maximizing the performance of simple, inexpensive filters. Especially when designing printed circuit boards, it pays to look ahead and provide via holes where capacitors or inductors may be added without redesigning the board. Actually, any-time you add capacitance or inductance to a circuit you are providing filtering. For example, bypassing and decoupling networks (Figures 5-1, 5-2, 6-1, 8-6) filter out power-supply noise, while spike-suppression networks (Figures 7-6, 7-7, 7-8, 8-3) filter out noise from inductors, motors, and rectifiers. We can also use small capacitors, small inductors, and ferrite beads to filter out noise on signal lines (Figures 5-3, 5-5, 5-6, 7-1, 7-2, 7-3) while preventing oscillations and other problems.

Filters attenuate noise signals three ways: (1) by shunting them to ground, (2) by reflecting them back toward the source, and (3) by converting them to heat. We usually refer to the first two types as "brute-force" filters because they aren't impedance-matched to the source or load. In most cases the source and load impedances are complex and frequency-dependent, so we try to provide a near short-circuit for desired signals, and a complete open (or a short to ground) for noise.

Figures 15-1(a) through (e) show some low-pass filters, which allow low-frequency signals to pass through unaffected but reflect high-frequency signals. These are the most commonly used noise filters. In many cases, a small capacitor or inductor (Figures 15-1(a), (b)) next to the noisy/sensitive circuit is all that is needed. For example, to prevent noise pickup on speaker wires, a 0.01 to 0.03 µF disc ceramic capacitor between each speaker terminal and the amplifier chassis usually suffices. To be most

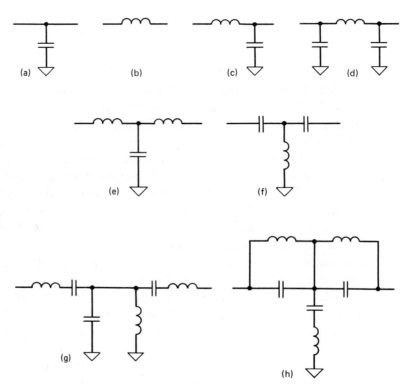

Figure 15-1 Discrete filters

effective, the self-resonant frequency of the capacitor/inductor should be in the middle of the interference band (see Figures 2-5 and 2-7). The "T-filter" (Figure 15-1(e)) provides the greatest protection against spikes and other transients.

Figure 15-1(f) shows a high-pass filter, which attenuates low-frequency signals and lets high-frequency signals pass through unaffected. This type of filter could be used with an FM radio, for example, to keep AM radio/CB broadcasts from causing audio-rectification problems. Figure 15-1(g) shows a bandpass filter, which passes only a narrow band of frequencies, and Figure 15-1(h) shows a notch filter, which blocks a narrow band of frequencies.

To reduce radio-frequency interference (RFI) we usually need to block a narrow band of frequencies. For this we use sharply tuned filters called "wavetraps." For noise signals below 40 MHz we can put a high-Q capacitor with long leads across the antenna terminals and trim or bend the leads to adjust the parasitic inductance and thus the resonant frequency. To block 40 to 100 MHz signals we can put an air-core RF choke in series with the signal wire, then stretch or squeeze the coil to adjust the resonant frequency. (*Note:* Use a grid-dip or gate-dip meter to set the wavetrap's resonant frequency slightly below the interference frequency, mount the wavetrap on the antenna terminals, and then adjust it to minimize interference.)

Above 100 MHz we can use "quarter-wave traps" and "half-wave traps." A quarter-wave trap is a piece of transmission line $\lambda/4$ meters long, open at the far end (Figures 15-2(b) and (d)). A half-wave trap is a piece of transmission line $\lambda/2$ meters long, shorted at the far end (Figures 15-2(a) and (c)). Figures 15-2(a) and (b) show wavetraps made from 300 Ω twinlead, hanging from a TV or radio's antenna terminals. Figures 15-2(c) and (d) show wavetraps made from coaxial cable (the one in Figure 15-2(c) has the shield soldered to the center conductor at the right end). In either case, we make the transmission line 50 to 100 mm longer than needed and install it on the TV or radio that is having problems. We test for interference, trim off ≈3 mm, test for interference again, etc. As long as we keep reducing the interference, we keep on trimming. The first time the interference gets worse, we stop—the wavetrap is as good as we can make it. For 300 Ω twinlead $\lambda \approx 2.49 \times 10^8/f$ meters, f in hertz. For RG-59U coaxial cable $\lambda \approx 2.37 \times 10^8/f$ meters. We can broaden the frequency range of a quarter-wave trap by terminating it with a ≈200 Ω resistor. Similarly, we can broaden the range of a half-wave trap by terminating it with a ≈5 Ω resistor.

Because inductive filters reflect high-frequency signals, they may set up standing waves on the wires and thus increase radiated noise. Inductive filters also tend to resonate, letting certain frequencies leak

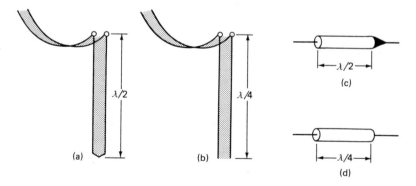

Figure 15-2   Wavetraps

through the filter without any attenuation. In fact, at resonance an inductive filter may double the noise voltage and *increase the noise problems*!

Lossy filters avoid these problems by turning unwanted signals into heat. The simplest lossy filter is a ferrite bead on a wire (Figure 15-3(a)), which acts like a short-circuit at low frequencies and like a 50–200 Ω resistor between 1 MHz and 100 MHz. Manganese-zinc ferrites work best up to 40 MHz, medium-permeability nickel-zinc ferrites are good up to 200 MHz, and low-permeability nickel-zinc ferrites are good above 200 MHz. We can increase the impedance by using long fat beads, by putting several beads in series (Figure 15-3(b)), or by running the wire through the bead several times (Figure 15-3(c)). Six-hole beads are convenient because we can run $2\frac{1}{2}$ turns of wire through the bead for 0.5–1 MHz noise, $1\frac{1}{2}$ turns of wire for 1–10 MHz noise, and run the wire straight through the bead for ≥ 10 MHz noise. Ferrite beads are also nice for eliminating parasitic oscillation, because slipping a bead onto the base or gate lead of a transistor reduces the high-frequency feedback without affecting the low-frequency operation of the circuit.

Ferrite beads are usually good up to 5 A DC, but we can raise this limit by gently cracking the beads, forming small air gaps. A very effective filtering arrangement is to use feedthrough capacitors and ferrite beads on all the signal and power lines entering an enclosure. If RFI enters a TV or radio through the power cord, try connecting a 0.01 μF 1.4 kV disk ceramic capacitor between the hot wire and the chassis, and another one between the neutral wire and the chassis (with short leads), and wrap the power cord around a ≈ 13 mm-diameter ferrite antenna rod.

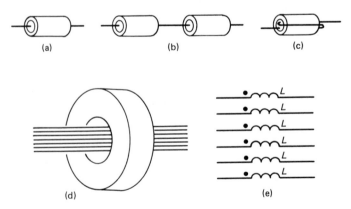

Figure 15-3   Ferrite bead/ferrite core filters

If RFI enters a TV or radio through a coaxial cable, run the cable through a cardboard tube, and stuff the tube full of steel wool.

If you have a cable with many signals to be filtered, run the cable through a ferrite core (or clamp a two-piece ferrite core around the cable) to form a common-mode choke (Figure 15-3(d)). Figure 15-3(e) shows the equivalent circuit for this device. Because every wire has L henries inductance, and every pair of wires has L henries mutual inductance, DC signals and differential-mode signals (equal currents leaving and returning on the cable) see a short circuit, while common-mode signals (current returning on another wire) are blocked by L henries inductance.

When designing filters, we must choose components with adequate AC-voltage, AC-current, and temperature ratings. Especially in power-line filters, the capacitors must withstand high ripple currents and the inductors must be able to safely carry at least twice the filter's rated current. These inductors may be wound in several sections to reduce their distributed capacitance and avoid unwanted resonances. The leads of all the components should be kept as short as possible to reduce noise emissions and noise pickup. When using rolled-foil capacitors (extended-foil construction is best), the outside foil (the end with the band) should go to ground. Standard rolled-foil capacitors are good to ≈ 20 MHz, while mica and standard ceramic capacitors are good to ≈ 200 MHz. The capacitors should have a working voltage of at least twice the peak input voltage.

Figure 15-4 shows a glitch-filter, to filter out voltage spikes on digital input lines. An RC-time constant of about 10 µs usually works well. Figure 15-5 shows a noise filter for an SCR or triac. The 0.1 µF capacitor should be rated for $\geq 1.67V_S$, where V_S is the root-mean-square (RMS) voltage of the power source; a light dimmer operating on 120 V AC would have a capacitor rated for at least 200 V. Figure 15-6 shows two filters for DC motors. $C1$ and $C2$ are 0.01 to 0.1 µF ceramic capacitors, and $L1$ and $L2$ are RF chokes, while $C3$ and $C4$ are feedthrough capacitors, and $L3$ and $L4$ are ferrite beads. The capacitors should be connected to the shield around the motor or to the motor casing.

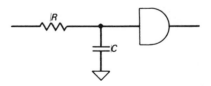

Figure 15-4 Glitch filter for digital signals

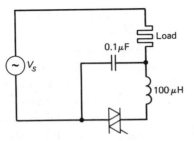

Figure 15-5   Filter for SCR/triac noise

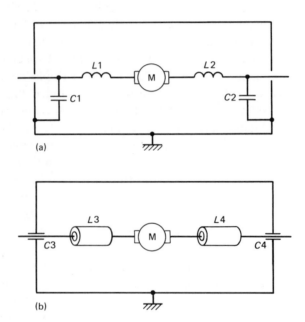

Figure 15-6   Filters for DC motors

In late 1983 we started having problems with some microprocessors, trying to test them at 141 kHz when their specifications called for 6 to 12 MHz clocks. So we developed a new test, which used special ''RAM Boards'' in the test fixtures to let the microprocessors run at 12 MHz while the tester ran at 200 kHz. The new test worked beautifully in the first three test fixtures. The fourth test fixture had problems—one side would fail 80% of the time, and the other side failed 30% of the time, with random errors. I spent two days chasing this problem and finally discovered that the ALE signal on the RAM Board had a small glitch. On this test fixture, the glitch would scramble the data to the microprocessor about 0.001% of the time, causing a failure. I tried bypassing ALE to ground with some small ceramic capacitors and found that 47 pF to 390 pF capacitors eliminated the glitch without causing any side-effects. My final solution, tacking a 200 pF capacitor between ALE and ground on the RAM Board, has worked like a charm on every test fixture we have built since then.

RECOMMENDED READING

Consumer Electronics Systems Technician Interference Handbook—TV Interference. Washington, D.C.: Consumer Electronics Group/Electronic Industries Association, no date.

Handbook on Radio Frequency Interference, Vol. 3. Wheaton, MD: Frederick Research Corp., 1962.

DARR, JACK, *Eliminating Man-Made Interference*. Indianapolis: Howard W. Sams & Co., Inc., 1960.

OTT, HENRY W., *Noise Reduction Techniques in Electronic Systems*. New York: John Wiley & Sons, 1976.

16

FINDING AND FIXING NOISE PROBLEMS

Noise problems fall into two categories: hard failures and soft failures. Hard failures are consistent—the system fails the same way every time. This lets us identify the problem with relative ease. We may have difficulty correcting the problem, but when we do, the hard failure disappears.

Soft failures are intermittent—sometimes the system works, and sometimes it fails. This on-again/off-again behavior makes problem identification much more difficult. Many times the system is acting up, and we seem to be getting closer to the problem, when suddenly the system starts working correctly and we don't know what we did! This can be very frustrating, because nine times out of ten the system will start failing again as soon as our boss enters the room. And even when we think we have found and cured the problem, we can't be sure.

As a designer of automatic test equipment, I have faced this dilemma many times. In essence, every time we connect a new board/subassembly/product to a tester, we create a brand new system. If there are any tester-product combinations that won't work, sooner or later they will start interfering with production, and the production line manager will tell me to fix the tester! (Hence my emphasis on Taylor Worst-Case Design—it helps keep me out of trouble.)

Through long, hard experience I have developed the following procedure for finding and fixing soft failures:

Step 1: Increase the number of failures.

Step 2: Eliminate as many factors as possible.

Step 3: Closely examine the remaining pieces. If something looks suspicious, try changing it to see what effect it has on the failures.

Step 4: Correct the defect and prove that the failures are gone.

To increase the number of failures, try to identify the combination that fails most often. Using this system as your test-bed, try varying supply voltages, signal levels, signal timing, input data, temperature, and anything else that might affect the number of failures. Record your experiments and their results to guide you through the next steps.

Eliminate everything that doesn't contribute to the failures. Try patching out code, eliminating tests, disconnecting cables, and removing subassemblies to find the smallest chunk of hardware and software causing the failure. If removing an "unrelated" piece of code or hardware makes a dramatic change in the system's behavior, you may have found the culprit. Now force the system to loop on the failing operation.

Examine the failures closely. What is being tested? How does it fail? When does it fail? Where does it fail? What else happens before, during, and after the failures? Are the failures mechanical, software, electrical, or electronic? Check everything you can, looking for anything out of the ordinary. Examine signals with a fast oscilloscope, starting with the failing signals and working toward the inputs. Check the power-, ground-, input-, and output-pins on each module in the path. Check for long rise and fall times, back-porching, overshoot, glitches, ringing, oscillations, and noise. What do the signals look like when the test fails, when the test passes, and how do they differ? How noisy are the power and ground lines? What happens when you touch a component pin with the scope probe or your finger? Get another person to look at the problem with you; they may spot something that you have overlooked.

Check ground noise with a 100-MHz oscilloscope, a CRT hood, a 10× probe, and a short ground lead. Attach the ground lead to one ground point and touch the 10× probe to another ground point to measure the maximum voltage between them. Check all critical grounds in the system, and spot-check other grounds. In digital circuitry, the ground noise should be less than 150 mV to 300 mV. If the ground noise exceeds 500 mV, you have probably found the culprit. Sketch the complete power/ground system, with all its interconnections, and look for daisy-chained grounds, ground loops, missing grounds, loose connections, etc. (see Chapter 10).

Check the power-supply voltage across each IC on the failing board(s) and on the backplane/cabling. Over 250 mV V_{cc}-to-ground noise on a board, or over 50 mV V_{cc}-to-ground noise on a backplane (20 mV for ECL) indicates a power-distribution/bypassing problem.

Check the signals going into ICs. If a signal rings on both the rising and falling edges, the ground system has too much inductance. You need to use capacitors with less inductance (*ESL*), lower the ground trace inductance, switch fewer signals at one time, or increase the load capacitance. If a signal rings only on the rising edges (ground-to-voltage) the circuit needs more bypass capacitance. Stretched-out rising edges, with normal falling edges, also indicate a need for better bypassing.

If an analog circuit is motorboating or oscillating, first check the bypassing networks. Try adding small capacitors between +inputs and ground, and across feedback resistors, to see if they help. Also try lower-value resistors in the networks going to +inputs of opamps. Check high-power amplifiers for parasitic oscillations with a neon lamp—if the lamp glows yellow, you have low-frequency oscillations present, if it glows violet, high-frequency oscillations.

If a problem seems to vary throughout the day, power-line transients may be the cause of the problem. Most of the newer digital oscilloscopes can capture single transients and can be a tremendous help chasing these problems. Visually inspect the system, looking for high-power circuits close to sensitive circuits, high-voltage/high-current wires paralleling sensitive wires, missing or misplaced grounds and shields, etc. Try holding your hand near different devices; if the symptoms vary with the position of your hand, you probably have a shielding problem. A 50 Ω $\frac{1}{4}$ W resistor soldered to the end of some RG-58/U cable makes a good ''sniffer probe'' for an oscilloscope or a radio receiver, for locating leaky shields.

After finding the defect, devise a cure. A capacitor between a signal and ground can fix minor noise problems. Serious noise problems may require extensive changes to grounds, cables, and circuits, or the addition of shields and filters.

Modify the system, then reexamine the signals that you checked in step 3. If they all look good, run through the combinations that you tried in step 1 all over again. If the system always passes the test, the problem is solved. Otherwise, repeat the process to find the next defect.

If you are designing a product for sale, you can save a lot of time and money by testing the product during development. Electrostatic discharge (ESD) tests can easily identify weak grounds, noise-sensitive cables and printed circuit boards, power-distribution problems, and shielding and filtering problems, with a minimum investment in equipment. The book *Digital Design for Interference Specifications*, by R. Kenneth Keenan, tells how to set up a simple EMC lab for $1,500 to $5,000, and how to perform the most important EMC tests. Henry Ott recommends using a current probe to check the common-mode current on cables. For a system with 1-meter-long cables, $\geqslant$15 μA common-mode current will exceed the FCC Class B limits, and $\geqslant$5 μA common-mode current will exceed the Class A limits. (*Note:* If a product costs less than $5,000, the FCC will probably put it in Class A no matter what the manufacturer says.)

Using the procedures in Keenan's book, you can identify some noise problems by disconnecting cables from the system-under-test. If disconnecting a particular cable reduces the emissions (usually peaking in the 30–100 MHz band for 1 m-long cables), you have a differential-mode problem with a signal(s) in that cable. If the emissions are the same whether one or many cables are connected, you have a common-mode problem. You will need to increase the rise/fall time of signals, improve the ground system, add common-mode chokes, or shield the cables. If the emissions are unchanged with all cables disconnected, you have either a clock-loop area problem or a power/ground busing problem.

In summary, there is no magical technique for solving noise problems. Every noise problem is different, so you must pay close attention to the details of the system. Be skeptical—take nothing for granted. Keep asking yourself, "Is this what I expect to happen?" When you get bogged down, take a break. If you get a hunch, check it out. Ask another person to look at the problem with you; they may see things that you had missed, and their questions may suggest new ways to attack the problem. If all else fails, browse through this book looking for similar types of problems, their causes, and their solutions.

RECOMMENDED READING

DARR, JACK, *Eliminating Man-Made Interference*. Indianapolis: Howard W. Sams & Co., Inc., 1960.

KEENAN, R. KENNETH, *Digital Design for Interference Specifications*. Pinellas Park, FL: TKC, 1983.

MILLER, GARY E., "Noise—the Silent Killer," *ATE Seminar/Exhibit Proceedings*, June 1968, 11–21 to 11–27.

NELSON, WILLIAM R., *Interference Handbook*. Wilton, Conn.: Radio Publications, Inc., 1984.

APPENDIXES

APPENDIX A: TAYLOR WORST-CASE DESIGN

Taylor Worst-Case Design is a simple method for computing realistic operating limits for circuits—limits encompassing the operating life of every unit produced. This design method is based on Murphy's Law: "Anything that can go wrong, will go wrong." It assumes that the most critical component in a circuit will drift the most, and that all the other components in the circuit will assume their most troublesome values.

We compute the Taylor Worst-Case Limits for a circuit as follows:

1. Get the specifications for all the components. Try to get nominal values, purchase tolerances, and end-of-life (EOL) tolerances. If you can't get the EOL tolerances for a component, use the estimators in this appendix.

2. Determine the circuit's critical output(s): voltage, current, gain, frequency, resistance, or whatever.

3. Sketch the circuitry affecting the output. Write down the component names, nominal values, purchase tolerances, and EOL tolerances.

4. Determine the most critical *passive* component in the circuit. This is the component that *minimizes* the output when it varies from its purchase tolerance to its EOL tolerance. Study the circuit, component values, and component tolerances to find this component. If several components look equally likely, analyze the circuit trying each candidate in turn.

5. Assign the most critical component its EOL tolerance that minimizes the output.

6. Assign the other components their purchase tolerances that minimize the output.

7. Assign every power supply and input signal its specified value that minimizes the output.

8. Compute the circuit's output. This is the *Taylor Worst-Case Low Limit.*

9. Repeat steps 4 through 8, this time trying to *maximize* the output of the circuit, to compute the *Taylor Worst-Case High Limit. (Note:*

The most critical component for the low limit is usually also the most critical component for the high limit.)

Let's work an example. Figure A-1 shows a test circuit for a printer board. $I1$ represents one output of a printhead driver chip with on-board resistor $R1$, and resistor $R2$ simulates the printhead. We need to compute $V1$, the driver chip's output voltage. We find the following information in the component specifications:

$$36.1 \text{ mA} \leqslant I1 \leqslant 42.0 \text{ mA}$$
$$R1 = 270 \text{ } \Omega \pm 2.5\% \text{ purchase tolerance}$$
$$R2 = 301 \text{ } \Omega \pm 1.0\% \text{ purchase tolerance}$$

We couldn't find EOL tolerances for the resistors, so we use the "three-times rule" to get $R1$'s EOL tolerance $\approx \pm 7.5\%$ and $R2$'s EOL tolerance $\approx \pm 3.0\%$.

By inspection, $V1 = I1 (R1 + R2)$. $R1$ and $R2$ are equally important and have about the same value, but $R1$ has the wider EOL tolerance. So $R1$ is our most critical component. If we weren't sure, we would try $R1$ as the most critical component and then try $R2$ as the most critical component.

Let's start with the low limit. $I1$, $R1$, and $R2$ must all be small to minimize $V1$. $R1$ is our most critical component, so we use its -7.5% EOL tolerance, making $R1 = (270 \text{ } \Omega)(1 - 0.075) = 249.75 \text{ } \Omega$. $R2$ is not the most critical component, so we use its -1.0% purchase tolerance, making $R2 = (301 \text{ } \Omega)(1 - 0.010) = 297.99 \text{ } \Omega$. $I1$ is not a passive component, so we use its minimum specified value of $I1 = 36.1 \text{ mA}$. This makes $V1 = (36.1 \text{ mA})(249.75 \text{ } \Omega + 297.99 \text{ } \Omega) = 19.773 \text{ V}$ our Taylor Worst-Case Low Limit for the circuit.

Now for the high limit. We want $I1$, $R1$, and $R2$ to be large. $R1$ is again the most critical component, so we use its $+7.5\%$ EOL limit to

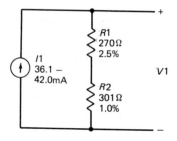

Figure A-1   Sample circuit for Taylor Worst-Case Design

get $R1 = (270\ \Omega)(1 + 0.075) = 290.25\ \Omega$. Similarly we use $R2$'s $+1.0\%$ purchase limit to get $R2 = (301\ \Omega)(1 + 0.010) = 304.01\ \Omega$. This time we want the maximum specified value for $I1$, so $I1 = 42.0$ mA. This makes $V1 = (42.0\ \text{mA})(290.25\ \Omega + 304.01\ \Omega) = 24.959$ V the Taylor Worst-Case High Limit for the circuit. So for this test, a $V1$ reading between 19.773 V and 24.959 V indicates a good board.

RULES OF THUMB FOR COMPONENT VALUES AND TOLERANCES

The values and tolerances needed for Taylor Worst-Case Design are often buried in the component specifications, or completely missing. The guidelines below are based on standard component specifications and IBM Lexington's experience in manufacturing and testing millions of printed circuit boards.

Bipolar Transistors (Non-Darlington)

$V_{be} \approx 0.35$ V to $V_{be}(\text{max})$
$V_{ce} \approx 0$ V to $V_{ce}(\text{sat max})$
$H_{fe} \geq 0.5\ H_{fe}(\text{min})$
Leakage resistance (C-E, C-B, E-B) $\geq 10\ \text{k}\Omega$

Capacitors (Electrolytic)

EOL tolerance $\approx$ purchase tolerance $\pm 40\%$

Capacitors (Nonelectrolytic)

EOL tolerance $\approx 3 \cdot$ purchase tolerance

Darlington Transistors

$V_{be} \approx 0.70$ V to $V_{be}(\text{max})$
$V_{ce} \approx 0.35$ V to $V_{ce}(\text{sat max})$

$R_{be} \geq V_{be}(\text{max})/I_b$, or $R_{be} \geq V_{be}(\text{max}) \cdot H_{fe}(\text{min})/I_c$
$H_{fe} \geq 0.5H_{fe}(\text{min})$
Leakage resistance (C-E, C-B, E-B) $\geq 10 \text{ k}\Omega$

Diodes

$V_f \approx 0.35 \text{ V}$ to $V_f(\text{max})$
Forward dynamic resistance $\approx 1 \text{ }\Omega$ to $12 \text{ }\Omega$
Leakage resistance $\geq 10 \text{ k}\Omega$ (below the breakdown voltage)

Field-Effect Transistors

$R_{ds} \approx 0 \text{ }\Omega$ to $R_{ds}(\text{max})$
$I_d \leq 10I_d(\text{max})$
Leakage resistance (D-S) $\geq 10 \text{ k}\Omega$

Inductors

EOL tolerance $\approx 3 \cdot$ purchase tolerance

Integrated Circuits

Resistor tolerance $\approx \pm 50\%$
Matching between resistors $\approx \pm 3\%$
Emitter-base breakdown (zener) voltage $\approx 6.6 \text{ V}$ to 7.4 V
Leakage of open-collector transistors $\approx -250 \text{ }\mu\text{A}$ to $250 \text{ }\mu\text{A}$
Delay times, propagation times, rise times, and fall times range from $0.5 \cdot$ typical to the maximum specified time
Other characteristics similar to discrete components

Relays

Contact resistance $\approx 0 \text{ }\Omega$ to $1.5 \text{ }\Omega$

Resistors

EOL tolerance $\approx 3 \cdot$ purchase tolerance

Spring-Loaded Probes

Contact resistance $\approx 0\ \Omega$ to $2.0\ \Omega$

Switches

Contact resistance $\approx 0\ \Omega$ to $1.5\ \Omega$

RECOMMENDED READING

BLAKESLEE, THOMAS R., *Digital Design with Standard MSI and LSI*. New York: John Wiley & Sons, 1975.
TAYLOR, NORMAN H., "Designing for Reliability," *Proceedings of the IRE*, 45:6 (June 1957), 811–822.

APPENDIX B: PROPERTIES OF TYPICAL CONDUCTORS

Table B-1 summarizes the important properties of conductors usually found in and around electronic systems. All good conductors have permittivity $\epsilon \approx 8.854$ pF/m ($\epsilon_r \approx 1$), but their other properties depend on their composition, processing, and environment. Proper heat treatment, for example, can increase permeability $\mu \approx 1.257\mu_r$ μH/m of some nickel alloys by a factor of 100.

The group numbers indicate the relative galvanic activity of the conductors, with group I being the most anodic (active) and group V the most cathodic (passive). Conductors within each group are compatible and will not corrode in the presence of moisture. If conductors from different groups are in contact *in the presence of moisture*, the more anodic conductor will corrode rapidly.

TABLE B-1 **Properties of Typical Conductors**

Conductor	Group	ρ, Resistivity (nΩ-m @ 20°C)	μ_r, Relative Permeability	Density (kg/m³)
Aluminum	II	27	1	2,700
Aluminum alloys	II	27–86	1	2,570–2,930
Beryllium copper	V	37–102	1	8,250
Brass	V	61–110	1	8,350–8,700
Bronze	V	91–212	1	7,570–8,850
Cadmium	II	73	1	8,640
Chromium	IV	132	1	7,100
Copper	V	17.2	1	8,960
Gold	V	22	1	19,300
Graphite	V	6,800–33,000	1	2,300–2,720
Iron	III	101	60–7,000	7,870
Lead	III	206	1	11,680
Magnesium	I	42	1	1,740
Magnesium alloys	I	50–143	1	1,750–1,870
Metglas[tm]	III	1,250	62,000–1,100,000	7,280
Monel	V	510–614	1	8,460–8,830
Mu-Metal[tm]	III	550–600	15,000–150,000	8,800
Nickel	IV	69	50–530	8,900
Nickel silver	IV	290	1	8,800
Permalloy[tm]	III	260–900	400–400,000	8,100–8,800
Platinum	V	106	1	21,450
Silver	V	16	1	10,500
Solder, tin-lead	III	150	1	8,890
Steel, mild	III	100–197	120–2,000	7,860
Steel, stainless	IV	560–780	1	7,730–7,960
Supermalloy[tm]	III	550–600	50,000–1,000,000	8,800
Tin	III	126	1	7,300
Titanium	V	540	1	4,500
Titanium alloys	V	482–1700	1	4,420–4,860
Zinc	II	60	1	7,140

RECOMMENDED READING

Bolz, Ray E., and George L. Tuve, eds., *Handbook of Tables for Applied Engineering Science*, 2nd ed. Cleveland: CRC Press, Inc., 1973.

Smithells, Colin J., *Metals Reference Book*, 5th ed. Boston: Butterworths, 1976.

APPENDIX C: PROPERTIES OF TYPICAL INSULATORS

Table C-1 summarizes the important properties of insulators usually found in and around electronic systems. All good insulators have permeability $\mu \approx 1.257 \ \mu H/m$ ($\mu_r \approx 1$), but their other properties depend on their

TABLE C-1   Properties of Typical Insulators

Material	ρ, Resistivity (Ω-m)	ϵ_r, Relative Permittivity	Dissipation Factor	Breakdown (MV/m)
ABS	1×10^{13}–1×10^{15}	2.4–5.0	.0030–.0150	11.8–17.7
ABS-polycarbonate	1×10^{13}–4×10^{14}	2.4–5.0	.0030–.0130	13.8–18.1
Acrylic	1×10^{12}–2×10^{14}	2.2–4.5	.0040–.0600	13.8–19.7
Air	1×10^9	1.00059	.0000	9.4
Alumina	1×10^9 –1×10^{12}	4.5–11.2	.0002–.0100	1.6–17.7
Barium titanate	1×10^6 –1×10^{13}	15.0–10,000	.0002–.0560	2.0–11.8
Beryllia	1×10^{12}–1×10^{15}	5.8–9.0	.0003–.0100	8.9–12.0
Epoxy	1×10^{10}–1×10^{12}	2.0–5.0	.0010–.0500	11.8–15.7
G-10 glass/epoxy	1×10^9 –4×10^{15}	3.5–5.9	.0030–.0870	11.8–23.6
Glass, borosilicate	1×10^6 –1×10^{15}	3.5–6.8	.0006–.0050	13.2–19.7
Glass, soda-lime	1×10^4 –1×10^{10}	5.9–8.3	.0010–.0110	0.8–60.3
Kapton	1×10^{14}–1×10^{15}	3.3–3.5	.0020–.0140	15.7–138
Kynar	3×10^{11}–2×10^{12}	6.4–8.4	.0180–.1700	10.2–78.7
Lexan	6×10^{13}–5×10^{14}	2.9–3.5	.0001–.0180	14.3–20.9
Mica	1×10^{12}–1×10^{15}	5.4–9.2	.0002–.0120	3.0–236
Mylar	1×10^{10}–1×10^{16}	2.8–7.3	.0020–.0600	10.8–138
Nylon	2×10^9 –1×10^{16}	3.1–7.6	.0090–.6000	11.8–110
Paper, impregnated	2×10^{11}–1×10^{17}	2.0–6.0	.0035–.0100	9.8–19.7
Phenolic	1×10^7 –1×10^{12}	4.0–21.0	.0050–.6400	4.7–15.7
Plexiglas$^{(tm)}$	1×10^{12}–2×10^{14}	2.2–4.5	.0040–.0600	13.8–19.7
Polyamide	2×10^9 –1×10^{16}	3.1–7.6	.0090–.6000	11.8–110
Polycarbonate	6×10^{13}–5×10^{14}	2.9–3.5	.0001–.0180	14.3–20.9
Polyester	1×10^{10}–1×10^{16}	2.8–7.3	.0020–.0600	10.8–138
Polyethylene	1×10^{13}–1×10^{16}	2.2–2.6	.0001–.0060	5.2–59.1
Polyimide	1×10^{14}–1×10^{15}	3.3–3.5	.0020–.0140	15.7–138
Polyphenelene oxide	1×10^{16}–1×10^{17}	2.6	.0003–.0009	15.7–21.7
Polypropylene	1×10^{13}–1×10^{15}	2.0–2.8	.0001–.0070	17.7–29.5
Polystyrene	1×10^{11}–1×10^{15}	2.4–4.8	.0001–.0050	7.9–125
Polyvinyl chloride	1×10^8 –1×10^{14}	2.8–9.0	.0060–.1500	8.9–44.7
Porcelain	1×10^8 –1×10^{13}	4.5–10.5	.0002–.0500	1.6–15.7
Quartz, fused	3×10^5 –1×10^{10}	3.7–3.9	.0001–.0009	16.1–39.4
Rubber	1×10^6 –1×10^{15}	2.1–13.0	.0023–.0700	5.9–41.3
Rubber, silicone	1×10^8 –1×10^{12}	3.2–6.3	.0020–.0200	7.5–18.3
Teflon fep	1×10^{16}–2×10^{18}	2.1	.0002–.0003	19.7–118
Teflon/glass	1×10^4 –1×10^{14}	2.2–5.0	.0003–.0040	9.8–63.0
Teflon tfe	1×10^{16}	2.0–2.2	.0001–.0002	5.6–33.5
Vacuum	1×10^9	1.00000	.0000	37.0–122
XXXP paper/phenolic	1×10^8	4.1–5.3	.0350–.0500	12.6–25.6

composition, processing, and environment. The proper impurities, for example, can increase permittivity $\epsilon \approx 8.854\epsilon_r$ pF/m of barium titanate by a factor of 600.

Most insulators are mixtures of compounds, so their electrical properties can vary widely. Designers can (1) design systems to tolerate these variations; (2) choose insulators that meet system requirements under worst-case conditions; or (3) control the insulators' composition, processing, and environment to reduce this variability.

RECOMMENDED READING

BOLZ, RAY E., and GEORGE L. TUVE, eds., *Handbook of Tables for Applied Engineering Science*, 2nd ed. Cleveland: CRC Press, Inc., 1973.

HARPER, CHARLES A., ed., *Handbook of Materials and Processes for Electronics*. New York: McGraw-Hill Book Co., 1970.

HARPER, CHARLES A., ed., *Handbook of Plastics and Elastomers*. New York: McGraw-Hill Book Co., 1975.

APPENDIX D: PROPERTIES OF ISOLATED WIRES

We can model any wire (PCB land, ground plane, etc.) as a resistance in series with an inductance, presenting impedance

$$| Z | = [R^2 + (2\pi f L)^2]^{1/2} \text{ ohms}$$

to signals of frequency f (hertz). R (ohms) depends on the wire's length, effective cross section, and resistivity. L (henries) depends on the wire's length, effective cross section, permeability, and overall shape. L depends also on the presence of nearby metal and on the path of the return current. (*Note:* The equations in this appendix assume that the wire is well away from the return-current path and large chunks of metal.)

As shown in Appendix H, the effective cross section depends on the skin depth.

$$\delta = \left(\frac{\rho}{\pi \mu_v \mu_r f} \right)^{1/2} \text{ meters,}$$

where ρ is the resistivity of the conductor $(\Omega - m)$, $\mu_v \approx 1.257 \ \mu H/m$, and $\mu_r \approx 1$ for nonmagnetic conductors. At x meters below the conductor's surface, the current density is

$$I(x) \approx I(0) \exp \left(\frac{-x}{\delta} \right) \ A/m^2,$$

where $I(0)$ is the current density at the surface (A/m^2). In essence, low-frequency signals flow almost uniformly across the entire cross section of the wire, but high-frequency signals act as though they flow in a skin only δ meters thick. This "skin effect" greatly increases the wire's resistance and slightly decreases its inductance.

Figure D-1 shows a straight, round wire of diameter D and length l (meters), made of a conductor with resistivity ρ $(\Omega\text{-}m)$ and relative permeability μ_r. For low frequencies, $f \leqslant f_b = 16\rho/(\pi\mu_v\mu_rD^2)$ hertz,

$$R \approx \frac{4\rho l}{\pi D^2} \ \text{ohms, and}$$

$$L \approx \frac{\mu_v l}{2\pi} \left[\ln \left(\frac{4l}{D} \right) + \left(\frac{\mu_r}{4} \right) - 1 + 0.389 \left(\frac{D}{l} \right) \right] \ \text{henries.}$$

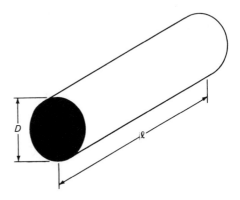

Figure D-1 A straight round wire

For high frequencies, $f > f_b$ hertz,

$$R \approx \frac{\rho l}{\pi D \delta} \text{ ohms, and}$$

$$L \approx \frac{\mu_v l}{2\pi} \left[\ln \left(\frac{4l}{D} \right) + \left(\frac{\mu_r \delta}{D} \right) - 1 + 0.5 \left(\frac{D}{l} \right) \right] \text{ henries.}$$

If the wire is bent into a circle, its low-frequency inductance is

$$L \approx \frac{\mu_v l}{2\pi} \left[\ln \left(\frac{8l}{\pi D} \right) + \left(\frac{\mu_r}{4} \right) - 2 \right] \text{ henries.}$$

Figure D-2 shows the total impedance of some 1 m-long round copper wires. Resistance dominates below 1 kHz, and inductance dominates above 100 kHz.

An N-gauge (AWG) solid wire has nominal diameter

$$D \approx 0.00825(0.89054)^N \text{ meters,}$$

with a diameter tolerance of $+0.5\%/-1.0\%$ or ± 2.54 μm, whichever is greater. An N-gauge (AWG) stranded wire has nominal diameter

$$D \approx 0.00971(0.89054)^N \text{ meters,}$$

with a diameter tolerance of $\pm 10\%$, depending on the number of strands.

Figure D-3 shows a straight PCB land, or rectangular strap, of width W, thickness $T \leq W$, and length l (meters), made of a conductor with resistivity ρ (Ω-m) and relative permeability μ_r. For low frequencies, $f \leq f_b = 4\rho(W + T)^2/(\pi\mu_v\mu_r W^2 T^2)$ hertz,

$$R \approx \frac{\rho l}{WT} \text{ ohms, and}$$

$$L \approx \frac{\mu_v l}{2\pi} \left[\ln \left(\frac{2l}{W + T} \right) + 0.5 + 0.2235 \left(\frac{W + T}{l} \right) \right] \text{ henries.}$$

For high frequencies, $f > f_b$ hertz,

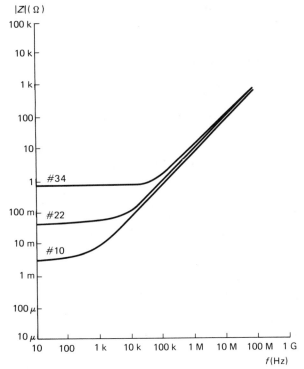

Figure D-2   Impedance of 1 m-long round copper wires

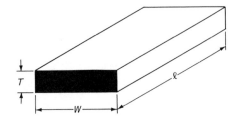

Figure D-3   A straight PCB land

$$R \approx \frac{\rho l}{2(W + T)\, \delta} \text{ ohms,}$$

$$GMD \approx (T + W)\left[0.2929 - 0.0698\left(\frac{W - T}{W + T}\right)^2\right] \text{ meters, and}$$

$$L \approx \frac{\mu_v l}{2\pi}\left[\ln\left(\frac{2l}{GMD}\right) - 1 + \left(\frac{GMD}{l}\right)\right] \text{ henries.}$$

Figure D-4 shows the total impedance of some copper PCB lands and straps. Resistance dominates below 10 kHz, and inductance above 100 kHz. A woven strap has slightly higher impedance than a solid strap of the same size, because of the weaving of the wires.

For copper-clad printed circuit boards, each "ounce" of copper is 35.6 μm ± 10% thick. The tolerance on land widths is approximately twice the copper thickness. The nominal thickness of N-gauge aluminum sheet is

$$T \approx 0.00838(0.89051)^N \text{ meters,}$$

while the nominal thickness of N-gauge steel sheet is

$$T \approx 0.0102(0.89156)^N \text{ meters.}$$

Figure D-5 shows a large ground plane of thickness T (meters), made of a conductor with resistivity ρ (Ω-m) and relative permeability

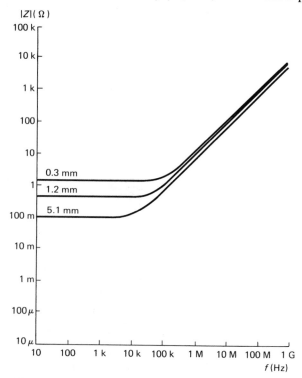

Figure D-4 Impedance of 1-m-long 36-μm-thick PCB lands

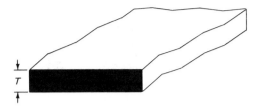

Figure D-5   A groundplane

μ_r. Letting $x = T/\delta$, we get

$$R = \frac{\rho\{1 - \exp(-x)[\cos(x) - \sin(x)]\}}{\delta\{1 - 2\exp(-x)\cos(x) + \exp(-2x)\}} \text{ ohms, and}$$

$$L = \frac{\rho\{1 - \exp(-x)[\cos(x) - \sin(x)]\}}{2\pi f\delta\{1 - 2\exp(-x)\cos(x) + \exp(-2x)\}} \text{ henries}$$

For low frequencies, $f \leq f_b = \rho/(\pi\mu_v\mu_r T^2)$ hertz,

$$R \approx \frac{\rho}{T} \text{ ohms, and}$$

$$L \approx \frac{\rho}{2\pi f T} \text{ henries.}$$

For high frequencies, $f > f_b$ hertz,

$$R \approx \frac{\rho}{\delta} \text{ ohms, and}$$

$$L \approx \frac{\rho}{2\pi f \delta} \text{ henries.}$$

Figure D-6 shows the total impedance between points on an infinite groundplane. The impedance is very low and is independent of the spacing between the points. For a finite groundplane, points in the middle see the calculated impedance, while points near the edges and corners see up to four times the calculated impedance.

Figure D-7 shows a straight thin-walled tube of outside diameter D, wall thickness $T \ll D$, and length l (meters), made of a conductor with resistivity ρ (Ω-m) and relative permeability μ_r. For low frequencies, $f \leq f_b = \rho/(\pi\mu_v\mu_r T^2)$ hertz,

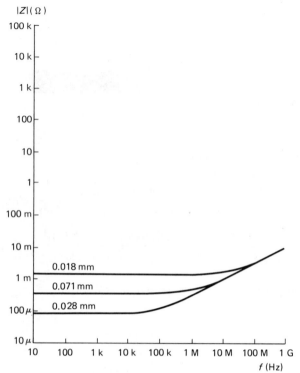

Figure D-6   Impedance of copper groundplanes

$$R \approx \frac{\rho l}{\pi D T} \text{ ohms, and}$$

$$L \approx \frac{\mu_v l}{2\pi} \left[\ln \left(\frac{4l}{D} \right) + \left(\frac{\mu_r T}{D} \right) - 1 + 0.5 \left(\frac{D}{l} \right) \right] \text{ henries.}$$

For high frequencies, $f > f_b$ hertz,

$$R \approx \frac{\rho l}{\pi D \delta} \text{ ohms, and}$$

$$L \approx \frac{\mu_v l}{2\pi} \left[\ln \left(\frac{4l}{D} \right) + \left(\frac{\mu_r \delta}{D} \right) - 1 + 0.5 \left(\frac{D}{l} \right) \right] \text{ henries.}$$

Figure D-8 shows the total impedance of some 1 m-long round copper tubes. Copper tubing is nearly ideal as a high-frequency conductor be-

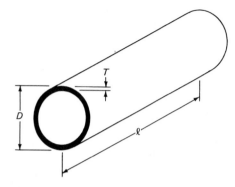

Figure D-7   A straight thin-walled tube

cause it is robust, readily available, and at high frequencies has low impedance for its weight.

Figure D-9 shows the impedances of typical wires and groundplanes made of mild steel, aluminum, and copper to illustrate the effects of the conductor on the wiring impedance. Notice the difference in the slope

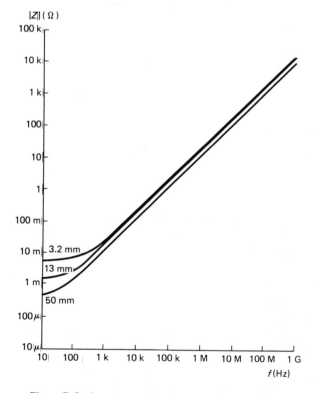

Figure D-8   Impedance of 1 m-long round copper tubes

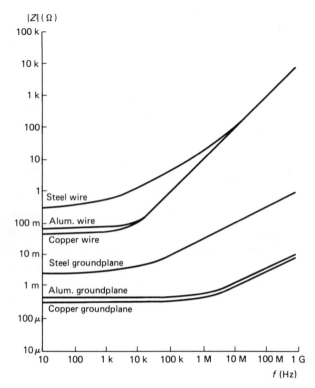

Figure D-9 Effect of materials on impedance

of the curves at high frequencies—a wire's impedance is proportional to frequency, while a groundplane's impedance is proportional to the square root of the frequency.

RECOMMENDED READING

GROVER, FREDERICK W., *Inductance Calculations*. Instrument Society of America, 1973.

RAMO, SIMON, JOHN R. WHINNERY, and THEODORE VAN DUZER, *Fields and Waves in Communication Electronics*. New York: John Wiley & Sons, Inc., 1965.

TERMAN, FREDERICK EMMONS, *Radio Engineers Handbook*. New York: McGraw-Hill Book Co., Inc., 1943.

WHITE, DONALD R. J., *EMI Control in the Design of Printed Circuit Boards and Backplanes*. Gainesville, VA: Don White Consultants, Inc., 1981.

APPENDIX E: PROPERTIES OF TRANSMISSION LINES

Whenever two or more parallel conductors carry equal currents in opposite directions, they form a transmission line. The shapes, sizes, positions, and permittivities of the conductors and insulators determine the line's electrical properties. If the conductors are uniform and have low resistance, the line will display nearly constant impedance from DC to 100 MHz and up.

We can model most system wiring as networks of lossless transmission lines. Wires in cables usually exhibit 50 Ω to 300 Ω impedance, while printed circuit board lands usually exhibit 50 Ω to 120 Ω impedance. Wirewrap wires, and other isolated wires, usually exhibit 100 Ω to 200 Ω impedance.

The circuit properties of a transmission line depend mainly on its length. A "short" transmission line has a propagation delay, from the driver to the farthest end, less than half the signal rise or fall time. Glitches return to the driver while the signal is still changing, and disappear in the rising/falling edge of the signal. We can analyze the circuitry using standard lumped-circuit techniques and can model the wiring capacitance as equivalent discrete capacitors if necessary.

A "long" transmission line has a propagation delay over half the signal rise/fall time. Glitches (overshooting, undershooting, spikes, and ringing) appear on the line after the rising/falling edge of the signal and may cause serious problems. We must use distributed circuit-analysis techniques to analyze the circuit, and must include time delays, line terminations, and line uniformity in the analysis (see Appendix F).

We can characterize a lossless transmission line by its average relative permeability $\mu_{r'}$, average relative permittivity $\epsilon_{r'}$, and shape factor sf. If the conductors are nonmagnetic and are well away from magnetic materials, $\mu_{r'} \approx 1$. If the fields between the conductors are mainly in air or vacuum (Figure E-1(a)), $\epsilon_{r'} \approx 1$. If the fields are mainly in an insulator (Figure E-1(b)), $\epsilon_{r'} \approx \epsilon_r$ of the insulator. If the fields are partly in air (vacuum) and partly in an insulator (Figure E-1(c)), $1 < \epsilon_{r'} < \epsilon_r$ of the insulator. A long, lossless transmission line has

$$C_u = \frac{\epsilon_v \epsilon_{r'}}{sf} \approx \frac{8.854 \epsilon_{r'}}{sf} \text{ pF/m capacitance,}$$

$$L_u = \mu_v \mu_{r'} sf \approx 1.257 \mu_{r'} sf \text{ } \mu\text{H/m inductance,}$$

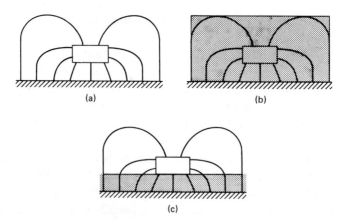

Figure E-1 Effect of insulators on average relative permittivity: (a) in air, (b) in thick insulator, (c) on thin insulator

$$t_u = (L_u C_u)^{1/2} \approx \frac{(\mu_{r'} \epsilon_{r'})^{1/2}}{2.998 \times 10^8} \text{ s/m propagation delay, and}$$

$$Zo = \left(\frac{L_u}{C_u}\right)^{1/2} \approx 376.7 \left(\frac{\mu_{r'}}{\epsilon_{r'}}\right)^{1/2} sf \ \Omega \text{ impedance.}$$

Vacuum and air have permittivity $\epsilon_v \approx 8.854$ pF/m and permeability $\mu_v \approx 1.257$ μH/m. If we have several nonmagnetic conductors with thin (or no) insulation forming a transmission line with impedance Zo (Ω) in air ($\mu_r \approx 1$, $\epsilon_r \approx 1$), we can rewrite these equations as

$$C_u \approx \left(\frac{3.34}{Zo}\right) \text{ nF/m capacitance,}$$

$$L_u \approx 3.34 Zo \text{ nH/m inductance, and}$$

$$t_u \approx 3.34 \text{ ns/m propagation delay.}$$

If the transmission line has thick insulation, and we are given capacitance C_u (F/m), we can write:

$$L_u \approx C_u Zo^2 \text{ H/m inductance, and}$$

$$t_u \approx C_u Zo \text{ s/m propagation delay.}$$

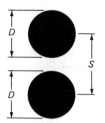

Figure E-2 Cross section of two-wire
transmission line

The shape factor of the transmission line, *sf*, depends only on the shapes, sizes, and positions of the conductors. Two-wire transmission line (Figure E-2) has shape factor

$$sf = \frac{1}{\pi} \ln \left[\frac{S}{D} + \left(\frac{S^2}{D^2} - 1 \right)^{1/2} \right]$$

when both wires have diameter D. When the wires have diameters $D1$ and $D2$, the shape factor becomes

$$sf = \frac{1}{2\pi} \ln [x + (x^2 - 1)^{1/2}]$$

where

$$x = \frac{4S^2 - D1^2 - D2^2}{2D1D2}$$

Twisting a two-wire transmission line to form a twisted-pair increases the average relative permittivity $\epsilon_{r'}$. For hard insulation with relative permittivity ϵ_r,

$$\epsilon_{r'} \approx 1 + (0.25 + 0.0004\Theta^2)(\epsilon_r - 1),$$

and for soft insulation (Teflon, polyvinyl chloride)

$$\epsilon_{r'} \approx 1 + (0.25 + 0.001\Theta^2)(\epsilon_r - 1),$$

where

$$\Theta = \arctan (\pi Sn) \text{ degrees},$$

and n is the number of twists per meter.

A wire near a ground plane (Figure E-3) has shape factor

$$sf = \frac{1}{2\pi} \ln \left[\frac{2S}{D} + \left(\frac{4S^2}{D^2} + 1 \right)^{1/2} \right].$$

A laminar bus (Figure E-4) has shape factor

$$sf \approx \frac{1}{\pi} \ln \left(\frac{4S}{W} + \frac{W}{2S} \right) \quad \text{for } W \le \frac{S}{2}, T << W,$$

$$sf \approx \frac{2}{\dfrac{2W}{S} + 2.42 - \dfrac{0.22S}{W} + \left(1 - \dfrac{S}{2W} \right)^6} \quad \text{for } W > \frac{S}{2}, T << W.$$

If an insulator separates the conductors (Figure E-1(c)),

$$\epsilon_{r'} \approx \frac{\epsilon_r + 1}{2} + \frac{\epsilon_r - 1}{\left(4 + 20 \dfrac{S}{W} \right)^{1/2}}.$$

Microstrip (Figure E-5) has shape factor

$$sf \approx \frac{1}{2\pi} \ln \left(\frac{8S}{W} + \frac{W}{4S} \right) \qquad \text{for } W \le S, T << W,$$

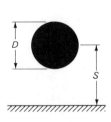

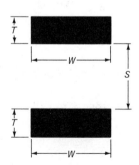

Figure E-3 Cross section of wire near a groundplane

Figure E-4 Cross section of a laminar bus

$$sf \approx \frac{1}{\dfrac{W}{S} + 2.42 - \dfrac{0.44S}{W} + \left(1 - \dfrac{S}{W}\right)^6} \qquad \text{for } W > S, \ T << W.$$

If an insulator separates the conductors (Figure E-1(c)),

$$\epsilon_{r'} \approx \frac{\epsilon_r + 1}{2} + \frac{\epsilon_r - 1}{\left(4 + 40\dfrac{S}{W}\right)^{1/2}}.$$

Coplanar lines (Figure E-6) have shape factor

$$sf \approx \frac{1}{\pi} \ln \left[2 \frac{\left(\dfrac{2W}{S} + 1\right)^{1/2} + 1}{\left(\dfrac{2W}{S} + 1\right)^{1/2} - 1} \right] \qquad \text{for } W \leq 2.414S, \ T << W,$$

$$sf \approx \frac{\pi}{4 \ln \left[2 \left(\dfrac{2W}{S} + 1\right)^{1/2} \right]} \qquad \text{for } W > 2.414S, \ T << W.$$

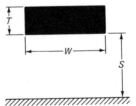

Figure E-5   Cross section of microstrip

Figure E-6   Cross section of coplanar lines

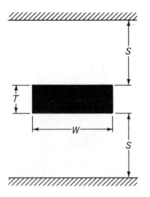

Figure E-7 Cross section of triplate stripline

If the lines are supported by an insulator on one side,

$$1 < \epsilon_{r'} < \frac{\epsilon_r + 1}{2}.$$

Triplate stripline (Figure E-7) has shape factor

$$\text{sf} \approx \frac{1}{2\pi} \ln \left[2\frac{\exp\left(\dfrac{\pi W}{4S}\right) + 1}{\exp\left(\dfrac{\pi W}{4S}\right) - 1} \right] \quad \text{for } W \leq 1.117S, \; T << S,$$

$$\text{sf} \approx \frac{1}{1.765 + \dfrac{2W}{S}} \quad \text{for } W > 1.117S, \; T << S.$$

Coaxial cable (Figure E-8) has shape factor

$$sf = \frac{1}{2\pi} \ln \left(\frac{B}{D}\right).$$

Shielded twisted-pair (Figure E-9) has shape factor

$$sf = \frac{1}{\pi} \ln \left(\frac{2S}{D} \frac{B^2 - S^2}{B^2 + S^2}\right).$$

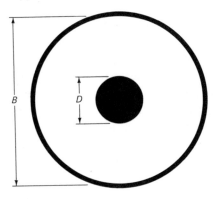

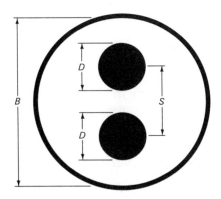

Figure E-8 Cross section of coaxial cable   **Figure E-9** Cross section of shielded
twisted-pair

Most of the formulas in this appendix have better than ±2% accuracy over their entire range. This is close enough for most engineering analyses: normal manufacturing tolerance on wire diameter is ±1% to ±10%. Commercial-grade printed circuit boards have ±5% to ±20% tolerance on the laminate thickness, ±10% tolerance on the copper thickness, and a land-width tolerance of approximately twice the copper thickness. Variations of ±5% in the relative permittivity of an insulator, over temperature and frequency, are also quite common.

RECOMMENDED READING

GUNSTON, M. A. R., *Microwave Transmission-Line Impedance Data*. New York: Van Nostrand Reinhold Co., 1972.

HILBERG, WOLFGANG, *Electrical Characteristics of Transmission Lines*. Dedham, MA: Artech House Books, 1979.

KEENAN, R. KENNETH, *Decoupling and Layout of Digital Printed Circuits*. Pinellas Park, FL: TKC, 1985.

LEFFERSON, PETER, "Twisted Magnet Wire Transmission Line," *IEEE Transactions on Parts, Hybrids, and Packaging*, PHP-7:4 (December 1971), 148–154.

SAAD, THEODORE S., ed., *Microwave Engineer's Handbook*, Vol. 1. Dedham, MA: Artech House, Inc., 1971.

APPENDIX F: REFLECTIONS ON TRANSMISSION LINES

At low frequencies most wires are electrically "short" ($l \leq 0.5t_r/t_u$ meters, $l \leq \lambda/2\pi$ meters) and just add capacitance to the signal nodes. But at high frequencies these same wires become electrically "long"; sharp bends in the wires, connectors, connections to drivers and receivers, and terminations at the ends of the wires create abrupt impedance changes which mutilate high-frequency signals on the wires. If a signal wire and the signal return path are approximately parallel, we can use transmission-line theory to determine the effects of these discontinuities on our signals.

Figure F-1 shows the model for a long transmission line connecting a driver, on the left, and a receiver, on the right. The driver has open-circuit output voltage V_S and output resistance R_S. The receiver has input resistance R_L and input capacitance C_L. The transmission line has overall capacitance C, inductance L, propagation delay t_p, and impedance Zo', where

$$C = C_u l + C_L \text{ farads,}$$

$$L = L_u l \text{ henries,}$$

$$t_p = (LC)^{1/2} = t_u l \left(1 + \frac{C_L}{C_u l}\right)^{1/2} \text{ seconds, and}$$

$$Zo' = \left(\frac{L}{C}\right)^{1/2} = Zo \left(\frac{C_u l}{C_u l + C_L}\right)^{1/2} \text{ ohms,}$$

and C_u, L_u, t_u, and Zo are the characteristic values for the infinitely long, lossless transmission line.

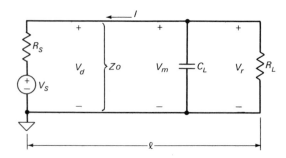

Figure F-1   Model for a long transmission line

Assume that V_S has been sitting at voltage V_i for a long time. Letting V_d be the voltage at the driver, V_m the voltage at the middle of the transmission line, and V_r the voltage at the receiver, we have

$$V_d = V_m = V_r = V_i \frac{R_L}{R_S + R_L} \text{ volts, } t \leq 0 \text{ seconds.}$$

Now let V_S suddenly change to voltage V_f at $t = 0$. This voltage change gets divided between R_S and Zo, putting voltage step

$$\Delta V1 = (V_f - V_i) \frac{Zo}{R_S + Zo} \text{ volts}$$

on the line, and the voltage at the driver becomes

$$V_d = V_i \frac{R_L}{R_S + R_L} + (V_f - V_i) \frac{Zo}{R_S + Zo} \text{ volts, } \qquad t > 0 \text{ seconds.}$$

The voltage step reaches the midpoint at $t \approx 0.5t_p$, so

$$V_m = V_i \frac{R_L}{R_S + R_L} + (V_f - V_i) \frac{Zo}{R_S + Zo} \text{ volts, } \qquad t > \frac{t_p}{2} \text{ seconds.}$$

At $t = t_p$ the voltage step reaches the receiver. It reflects off the parallel combination of R_L and C_L, with reflection coefficient $(R_L - Zo')/(R_L + Zo')$, creating a new voltage step

$$\Delta V2 = (V_f - V_i) \frac{Zo}{R_S + Zo} \frac{R_L - Zo'}{R_L + Zo'}$$

making

$$V_r = V_i \frac{R_L}{R_S + R_L} + (V_f - V_i) \frac{Zo}{R_S + Zo} \frac{2R_L + Zo' - Zo}{R_L + Zo'} \text{ volts}$$

for $t > t_p$ seconds. If $R_L \neq Zo'$, this new voltage step propagates back toward the driver, reflects off R_S in the same way, and on and on it goes. Given enough time, the voltage on the transmission line will settle down to

$$V_d = V_m = V_r = V_f \frac{R_L}{R_S + R_L} \text{ volts.}$$

Figure F-2 shows typical waveforms for V_d, V_m, and V_r for a positive step and the nine possible combinations of end terminations. A negative step would have the same waveforms, only inverted. Notice that when $R_L = Zo'$, the signal settles to its final value, all along the line, in time t_p. When $R_L \neq Zo'$, but $R_S = Zo$, the signal takes $2t_p$ to settle. And when $R_L \neq Zo'$ and $R_S \neq Zo$, the signal takes even longer to settle.

The procedure described above is limited to linear terminations and is rather tedious. If we need only $\pm 2\%$ or so accuracy, or need to analyze transmission lines with nonlinear terminations, we can solve these equations graphically using Bergeron diagrams.

Strictly speaking, Bergeron diagrams apply only to uniform transmission lines with lumped discontinuities at their ends, like the model in Figure F-1. But this fits most of the circuits we encounter in practice. The usual complication is multiple drivers and receivers. If several drivers

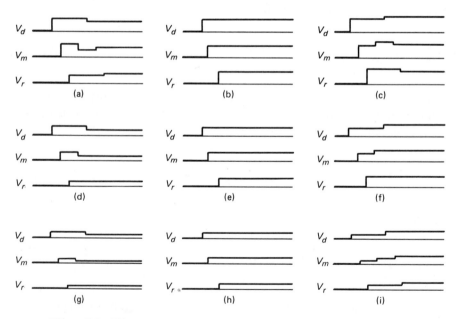

Figure F-2 Effects of terminations on the waveforms on long transmission lines: (a) $R_S < Zo$, $R_L < Zo'$, (b) $R_S < Zo$, $R_L = Zo'$, (c) $R_S < Zo$, $R_L > Zo'$, (d) $R_S = Zo$, $R_L < Zo'$, (e) $R_S = Zo$, $R_L = Zo'$, (f) $R_S = Zo$, $R_L > Zo'$, (g) $R_S > Zo$, $R_L < Zo'$, (h) $R_S > Zo$, $R_L = Zo'$, (i) $R_S > Zo$, $R_L > Zo'$

or receivers are connected to an end of the transmission line, we can just lump them together for the analysis. If one or more receivers connect to the middle of the line we can ignore them if $R_{in} \gg Z_O$ and $C_{in} \ll C_u l$, or include them as part of Z_O. If the driver is in the middle of the line its output resistance sees a transmission-line impedance of $Z_O/2$, which reduces the initial voltage jump, but the rest of the analysis proceeds normally.

Let's analyze a "1" to "0" transition for the circuit in Figure F-3(a), using a Bergeron diagram.

1. We start by marking the schematic with the current arrow and the voltage polarity. I is the current going *into* the driver, and V is the voltage on the signal line, referenced to signal ground.

2. Draw and label the axes for the Bergeron diagram ($V–I$ graph, Figure F-3(b)). $+ V$ goes at the top and $+I$ on the right. To make plotting easier, we can scale the voltage and current axes so that a 45-degree line represents Z_O ohms impedance. (*Note:* All the Bergeron diagrams in this appendix are scaled to $Z_O = 100\ \Omega$, the nominal impedance of readily available twisted-pair and ribbon cables.)

3. Draw and label voltage-time ($V–t$) graphs for the driver and receiver (Figure F-3(c)). $+ V$ goes on the top and $+t$ on the right. Label the t-axis in t_p (propagation time) units; for this example, $t_p \approx 50$ ns.

4. On the $V–I$ graph, plot and label "0-drive" and "1-drive" loadlines for the driver. These will have *positive* slopes. Also plot and label the loadline for the receiver. This will have a *negative* slope.

5. Find the intersection of the receiver loadline and the initial drive loadline. This is the initial voltage on the transmission line (point A on Figure F-3(b)).

6. From this point, draw a line with a scaled slope of $-Z_O$ ohms to the final drive loadline. This is the new drive voltage (point B) and represents the signal propagating to the receiver.

7. From this point, draw a line with scaled slope of $+Z_O$ ohms to the receiver loadline. This is the new receiver voltage (point C) and represents the signal propagating back to the driver.

8. Repeat steps 6 and 7 until you reach the intersection of the receiver loadline and the final drive loadline. Label these points D, E, F, etc.

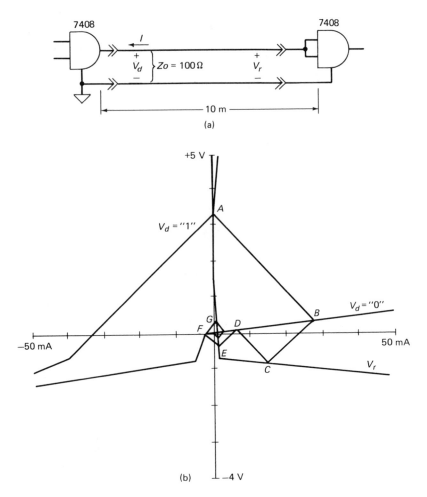

Figure F-3 Long transmission line with nonlinear terminations: (a) schematic, (b) Bergeron diagram, (c) waveforms

9. Plot these voltages on the *V-t* graphs. The driver sees voltage *A* for $t < 0$, voltage *B* for $0 \leq t < 2t_p$, voltage *D* for $2t_p \leq t < 4t_p$, etc. The receiver sees voltage *A* for $t < t_p$, voltage *C* for $t_p \leq t < 3t_p$, voltage *E* for $3t_p \leq t < 5t_p$, etc.

10. Study the waveforms. If we were using MOS ICs, and overshoot kicked a signal above V_{cc}, we could destroy a chip. If the signal rings or undershoots at the receiver, and gets between the maximum "0"-level and the minimum "1"-level, we are almost guaranteed

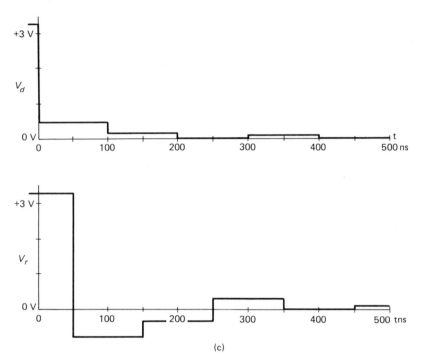

(c)

Figure F-3   Continued

noise problems in the circuit. Study the *V-I* plot to see how the dangerous voltage occurs, and what could be done to prevent it.

Figure F-4 shows driver and receiver loadlines for some common families of digital integrated circuits. The *V*-axes are marked in 1 V units and the *I*-axes in 10 mA units to allow plotting a 100 Ω transmission-line impedance as a 45-degree line.

Some general comments: The reflection at a termination is proportional to half the impedance mismatch. If R_L is within 20% of Zo' the reflected voltage step will be less than 10% of the incident voltage step. This is close enough for most digital circuits, but high-speed analog circuits and critical digital circuits may not tolerate this much degradation. Chapter 6 describes a practical method, using prototype hardware, to exactly match terminations to transmission lines.

For critical designs, or products that are going to be mass-produced, I strongly recommend using Taylor Worst-Case Design on all long cables. Characterize the transmission lines by using the equations in Appendix

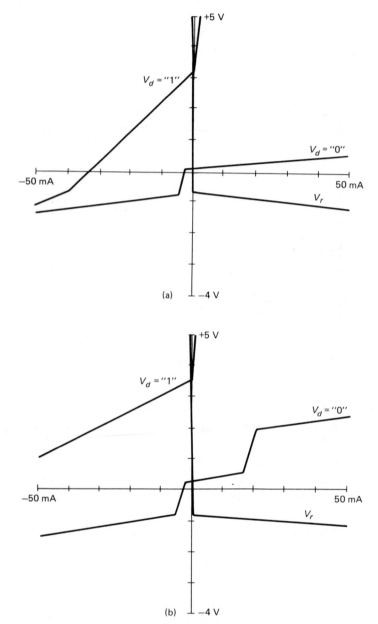

Figure F-4 Loadlines for common digital ICs: (a) 74xx TTL, (b) 74ALSxxx TTL, (c) 74ASxx TTL, (d) 74HCxx CMOS, (e) 74LSxx TTL, (f) 74Sxx TTL

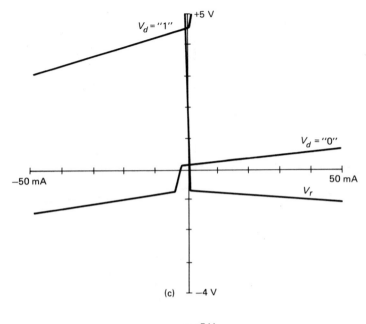

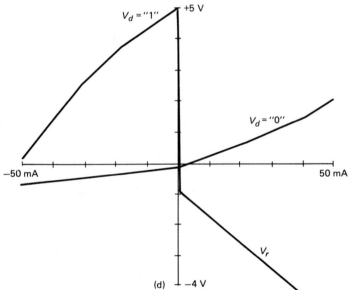

Figure F-4 Continued

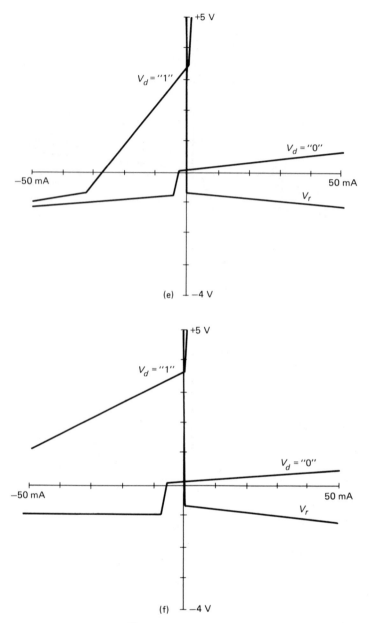

Figure F-4 Continued

E and the insulator data in Appendix C. Then draw Bergeron diagrams for the lowest- and highest-impedance transmission lines and for the lowest- and highest-resistance terminators (four graphs in all). Check the waveforms for overshoot, undershoot, ringing, glitches, and possible timing problems. If you see any problems, study the *V-I* graphs to see how the problems occur and what could be done to prevent them.

RECOMMENDED READING

BLAKESLEE, THOMAS R., *Digital Design with Standard MSI and LSI*. New York: John Wiley & Sons, 1975.

METZGER, GEORGES, and JEAN-PAUL VABRE (translated by Robert McDonough), *Transmission Lines with Pulse Excitation*. New York: Academic Press, 1969.

NORRIS, BRYAN, ed., *Digital Integrated Circuits and Operational Amplifier and Optoelectronic Circuit Design*. New York: McGraw-Hill Book Co., 1976.

SCARLETT, J. A., *Transistor-Transistor Logic and its Interconnections*. New York: Van Nostrand Reinhold Co., 1972.

SINGLETON, ROBERT S., "No Need to Juggle Equations to Find Reflection— Just Draw Three Lines," *Electronics*, 41:22 (October 29, 1968), 93–99.

APPENDIX G: CROSSTALK ON TRANSMISSION LINES

Crosstalk is the unwanted coupling of signals between circuits, mainly caused by running wires close to one another for long distances. Figure G-1 shows the general distributed-impedance crosstalk model, and Figures G-2 and G-3 show simplified lumped-impedance crosstalk models for small circuits.

In Figure G-1, the active circuit consists of a driver, V_S and R_S, driving load R_L through an *l*-meter-long transmission line. The active transmission line has total capacitance C_a farads, total inductance L_a henries, and impedance $Z_a = (L_a/C_a)^{1/2}$ ohms. The quiet circuit consists of an *l*-meter-long transmission line, parallel to the active transmission line, terminated by R_{ne} and R_{fe}. The quiet transmission line has total capacitance C_q farads, total inductance L_q henries, and impedance $Z_q = (L_q/C_q)^{1/2}$ ohms. The signal conductors of the active and quiet lines form a third transmission line with total mutual capacitance C_m farads, total mutual inductance L_m henries, and impedance $Z_m = (L_m/C_m)^{1/2}$ ohms.

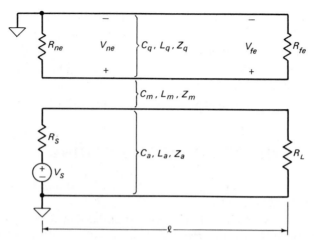

Figure G-1 Distributed-impedance crosstalk model

When studying crosstalk we are interested in the *disturbances from steady-state*. In the following analyses we ignore DC voltages on the signal lines and assume (1) the voltage source starts at 0 V, uniformly rises/falls to V_S volts over t_r seconds, and then stays there; (2) R_S includes the active driver's final output impedance; and (3) R_{ne} or R_{fe} includes the output impedance of the quiet line's driver. A complete crosstalk analysis includes

1. quiet line low and active line rising,
2. quiet line low and active line falling,
3. quiet line high and active line rising, and
4. quiet line high and active line falling.

The distributed-impedance crosstalk model can be greatly simplified if both transmission lines are short, with propagation delay $t_p \leq 0.5t_r$, so we don't have to worry about reflections. If both the active and the passive circuit have high impedances, $R_{ne} + R_{fe} > 376.7 \ \Omega$ and $R_S + R_L > 376.7 \ \Omega$, we can use the lumped-impedance capacitive-crosstalk model in Figure G-2. If both circuits have low impedances, $R_{ne} + R_{fe} < 376.7 \ \Omega$ and $R_S + R_L < 376.7 \ \Omega$, we can use the inductive-crosstalk model in Figure G-3.

Most analog and digital circuits on nongroundplane boards fit the lumped-impedance capacitive-crosstalk model. We derive this model from

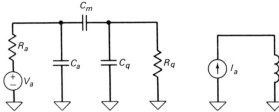

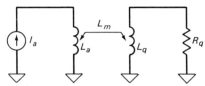

Figure G-2 Lumped-impedance capacitive-crosstalk model

Figure G-3 Lumped-impedance inductive-crosstalk model

Figure G-1 by ignoring L_a, L_m, and L_q and letting

$$V_a = \frac{V_S R_L}{R_S + R_L} \text{ volts,}$$

$$R_a = \frac{R_S R_L}{R_S + R_L} \text{ ohms, and}$$

$$R_q = \frac{R_{ne} R_{fe}}{R_{ne} + R_{fe}} \text{ ohms.}$$

Even with these simplifications the general solution for the crosstalk voltage is very complex, but three special cases have simple solutions. Let $t_a = R_a (C_a + C_m)$ be the active-loop time constant, and let $t_q = R_q (C_q + C_m)$ be the quiet-loop time constant. The maximum crosstalk voltage is

$$V_q \approx \frac{V_a R_q C_m}{t_r} \text{ volts} \qquad \text{when } t_a \ll t_r \text{ and } t_q \ll t_r,$$

$$V_q \approx \frac{V_a C_m}{C_q + C_m} \text{ volts} \qquad \text{when } t_a \ll t_q \text{ and } t_q \gg t_r, \text{ or}$$

$$V_q \approx \frac{V_a R_q C_m}{R_a (C_q + C_m)} \text{ volts} \qquad \text{when } t_a \gg t_q \text{ and } t_a \gg t_r.$$

Let's use this model to compute the maximum crosstalk on a 0.15 m × 0.20 m G-10 epoxy/glass board populated with 74LSxx TTL chips. Assuming x-y routing of the lands, the maximum land length is about

0.35 m. Lands on the board see average relative permittivity $\epsilon_{r'} \approx 4$, so the unit propagation delay is $t_u \approx 6.7$ ns/m. Multiplying 0.35 m by 6.7 ns/m, we get the propagation delay for a signal as $t_p \leq 2.35$ ns. This is less than half the typical rise and fall times for 74LSxx ICs ($t_r = 19$ ns, $t_f = 4.9$ ns from Table 3-1), so we can use a lumped-impedance model. 74LSxx ICs have voltage swing $V_S \approx 3.3$ V, output impedance $R_S \approx 120\ \Omega$ (high) or $31\ \Omega$ (low), and input impedance $R_L \approx 20$ kΩ. With a fanout of 1 to 10, the circuit impedances ($R_S + R_L$, and $R_{ne} + R_{fe}$) are

$120\ \Omega + 20$ k$\Omega = 20{,}120\ \Omega$ for a fanout of 1 in the high state, or

$$120\ \Omega + \frac{20\ \text{k}\Omega}{10} = 2120\ \Omega \text{ for a fanout of 10;}$$

$31\ \Omega + 20$ k$\Omega = 20{,}031\ \Omega$ for a fanout of 1 in the low state, or

$$31\ \Omega + \frac{20\ \text{k}\Omega}{10} = 2031\ \Omega \text{ for a fanout of 10.}$$

All these impedances are above 376.7 Ω, so we can use the capacitive-coupling model.

Let $V_S \approx 3.3$ V, $31\ \Omega \leq R_S \leq 120\ \Omega$, 2 k$\Omega \leq R_L \leq 20$ kΩ, $31\ \Omega \leq R_{ne} \leq 120\ \Omega$, and 2 k$\Omega \leq R_{fe} \leq 20$ kΩ (R_{ne} and R_{fe} may be swapped without affecting the results). Substituting these values into the equations for the capacitive-crosstalk model, we get 3.11 V $\leq V_a \leq$ 3.32 V, 30.5 $\Omega \leq R_a \leq 119.3\ \Omega$, and 30.5 $\Omega \leq R_q \leq 119.3\ \Omega$. Let's assume $C_a \approx C_q \approx 15$ pF and $C_m \approx 2$ pF. This gives us 0.52 ns $\leq t_a \leq$ 2.03 ns and 0.52 ns $\leq t_q \leq$ 2.03 ns. Because $t_a \ll t_r$ and $t_q \ll t_r$, we have the first special case, so the maximum crosstalk voltage is

$$V_q \approx \frac{V_a R_q C_m}{t_r} = \frac{(3.3\ \text{V})(119.3\ \Omega)(2\ \text{pF})}{4.9\ \text{ns}} = 0.16 \text{ volt.}$$

This is well within the noise margin (*NML* = 0.55 V, *NMH* = 1.4 V) for 74LSxx TTL, so we don't need to worry about crosstalk on this board.

High-current signals such as clocks, solenoid drives, and motor drives can cause inductive crosstalk. If the transmission lines are short, $t_p \leq 0.5t_r$, and both the active and the quiet circuit have impedances under 376.7 Ω, we can use the lumped-impedance inductive-crosstalk model in Figure G-3. This is derived from the general model in Figure G-1 by ignoring C_a, C_m, and C_q. The maximum crosstalk voltage is

$$V_q \approx \frac{I_a L_m}{t_r} \approx \frac{V_S L_m}{(R_S + R_L)t_r} \text{ volts.}$$

The hard part is computing L_m. Appendix E covers a number of types of transmission lines, which may help. Another common arrangement is two round wires sharing a common groundplane, shown in Figure G-4. These wires have mutual inductance of

$$L_m = \frac{\mu_v \mu_r l}{2\pi}$$

$$\cdot \frac{\ln\left(\frac{4H1}{D1}\right) \ln\left(\frac{4H2}{D2}\right) - \left[\ln\left[\left(1 + \frac{4H1H2}{S^2}\right)^{1/2}\right]\right]^2}{\ln\left(1 + \frac{4H1H2}{S^2}\right)} \text{ henries}$$

and mutual capacitance of

$$C_m = \frac{2\pi\epsilon_v\epsilon_r l \ \ln\left(1 + \frac{4H1H2}{S^2}\right)}{\ln\left(\frac{4H1}{D1}\right) \ln\left(\frac{4H2}{D2}\right) - \left[\ln\left[\left(1 + \frac{4H1H2}{S^2}\right)^{1/2}\right]\right]^2} \text{ farads.}$$

If the signal propagation delay exceeds half the signal rise or fall time, we are forced to use the distributed-impedance crosstalk model in

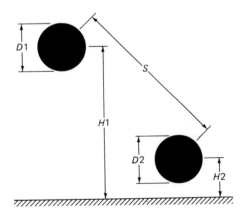

Figure G-4 Two wires sharing a common
groundplane

Figure G-1. This requires a five-step process: (1) Sketch V_{bc}, the backward crosstalk induced in the near end of the quiet circuit. (2) Sketch V_{fc}, the forward crosstalk induced in the far end of the quiet circuit. (3) Sketch the waveforms generated by V_{bc} propagating along the quiet line and reflecting off R_{ne} and R_{fe}. (4) Sketch the waveforms generated by V_{fc} propagating along the quiet line and reflecting off R_{fe} and R_{ne}. (5) Sketch the sum of the waveforms from steps 3 and 4.

Figure G-5 shows V_{bc} and V_{fc} induced in a long quiet transmission line, $(L_aC_a)^{1/2} + (L_qC_q)^{1/2} > t_r$, by a ramp of V_S volts in t_r seconds on the active line. The maximum backward crosstalk is

$$V_{bc} = \left(\frac{L_m}{2Z_a} + \frac{C_mZ_q}{2} \right) \left(\frac{V_S}{(L_aC_a)^{1/2} + (L_qC_q)^{1/2}} \right) \text{volts.}$$

The maximum forward crosstalk is

$$V_{fc} = - \left(\frac{L_m}{2Z_a} - \frac{C_mZ_q}{2} \right) \left(\frac{V_S}{t_r} \right) \text{volts.}$$

Figure G-6 shows V_{bc} and V_{fc} induced in a short quiet transmission line, $(L_aC_a)^{1/2} + (L_qC_q)^{1/2} \leq t_r$, by a ramp of V_S volts in t_r seconds on

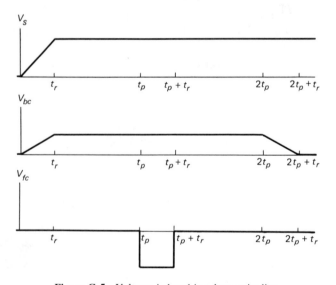

Figure G-5   Voltages induced in a long quiet line

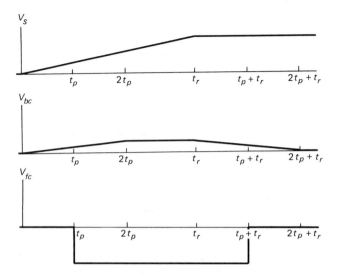

Figure G-6   Voltages induced in a short quiet line

the active line. The maximum backward crosstalk is

$$V_{bc} = \left(\frac{L_m}{2Z_a} + \frac{C_m Z_q}{2}\right)\left(\frac{V_S}{t_r}\right) \text{ volts.}$$

The maximum forward crosstalk is

$$V_{fc} = -\left(\frac{L_m}{2Z_a} - \frac{C_m Z_q}{2}\right)\left(\frac{V_S}{t_r}\right) \text{ volts.}$$

If the conductors are in a homogeneous insulator, $V_{fc} = 0$ V.

Figure G-7 shows the crosstalk analysis of two long lines, driven from the same end and terminated in their characteristic impedance, when the active line goes from low to high (Figure G-7(a)). The assumptions are $R_S \approx R_{ne} \approx 0\ \Omega$, $R_L \approx Z_a$, and $R_{fe} \approx Z_q$. The backward crosstalk (Figure G-7(b)) gets inverted by R_{ne}, propagates down the line, and then disappears into R_{fe} (Figure G-7(c)). The forward crosstalk (Figure G-7(d)) just disappears into R_{fe} (Figure G-7(e)). By summing the waveforms in Figure G-7(c) and (e), we get the actual waveforms that will appear on the quiet line (Figure G-7(f)).

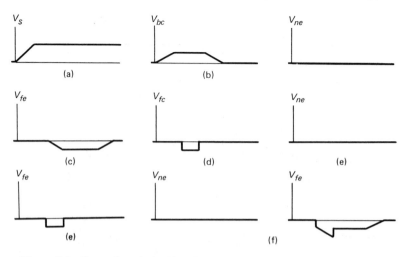

Figure G-7 Crosstalk analysis of long lines terminated in their characteristic impedance: (a) active signal, (b) induced near-end, (c) reflected near-end, (d) induced far-
end, (e) reflected far-end, (f) composite signal

Figure G-8 shows how different types of terminations on the quiet
line affect the crosstalk waveforms with V_S going from low to high. If
V_S goes from high to low, these waveforms will be inverted.

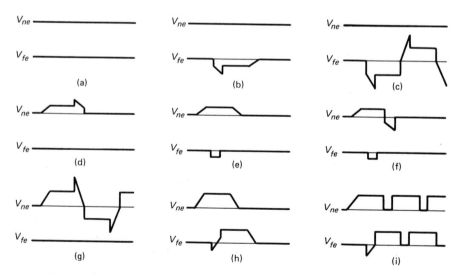

Figure G-8 Effect of quiet-line terminations on crosstalk waveforms: (a) $R_{ne} = 0$,
$R_{fe} = 0$, (b) $R_{ne} = 0$, $R_{fe} = Zo$, (c) $R_{ne} = 0$, $R_{fe} = \infty$, (d) $R_{ne} = Zo$, $R_{fe} = 0$,
(e) $R_{ne} = Zo$, $R_{fe} = Zo$, (f) $R_{ne} = Zo$, $R_{fe} = \infty$, (g) $R_{ne} = \infty$, $R_{fe} = 0$, (h) $R_{ne} = \infty$, $R_{fe} = Zo$, (i) $R_{ne} = \infty$, $R_{fe} = \infty$

RECOMMENDED READING

BLOOD, WILLIAM R., Jr., *MECL System Design Handbook*, 4th ed. Phoenix, AZ: Motorola Semiconductor Products, 1983.

GRAY, HARRY J., *Digital Computer Engineering*. Englewood Cliffs, NJ: Prentice-Hall, 1963.

MOHR, RICHARD J., "Interference Coupling—Attack It Early," *EDN*, 14:13 (July 1, 1969), 33–41.

APPENDIX H: ELECTROMAGNETIC FIELDS

Maxwell's equations describe the propagation of electromagnetic fields. In source-free regions (no net positive or negative charge), and using spherical coordinates, Maxwell's equations can be written:

$$\frac{\partial}{\partial \theta} (\sin \theta \, E_\varphi) - \frac{\partial E_\theta}{\partial_\varphi} = -j2\pi f\mu r \sin \theta H_r,$$

$$\frac{\partial}{\partial \theta} (\sin \theta \, H_\varphi) - \frac{\partial H_\theta}{\partial_\varphi} = \left(\frac{1}{\rho} + j2\pi f\epsilon\right) r \sin \theta \, E_r,$$

$$\frac{\partial}{\partial r} (rE_\theta) - \frac{\partial E_r}{\partial \theta} = -j2\pi f\mu rH_\varphi,$$

$$\frac{\partial}{\partial r} (rH_\theta) - \frac{\partial H_r}{\partial \theta} = \left(\frac{1}{\rho} + j2\pi f\epsilon\right) rE_\varphi,$$

$$\frac{\partial E_r}{\partial_\varphi} - \sin \theta \frac{\partial}{\partial r} (rE_\varphi) = -j2\pi f\mu r \sin \theta \, H_\theta, \text{ and}$$

$$\frac{\partial H_r}{\partial_\varphi} - \sin \theta \frac{\partial}{\partial r} (rH_\varphi) = \left(\frac{1}{\rho} + j2\pi f\epsilon\right) r \sin \theta \, E_\theta.$$

If a homogeneous medium has resistivity ρ (ohm-meters), permeability $\mu = \mu_v\mu_r \approx 1.257\mu_r$ μH/m, and permittivity $\epsilon = \epsilon_v\epsilon_r \approx 8.854\epsilon_r$ pF/m, we can define its *intrinsic impedance* as

$$\eta = \left(\frac{j2\pi f\mu\rho}{1 + j2\pi f\epsilon\rho}\right)^{1/2} \text{ ohms}$$

and its *intrinsic propagation constant* as

$$\sigma = \left[j2\pi f\mu \left(\frac{1}{\rho} + j2\pi f\epsilon \right) \right]^{1/2}$$

We often express σ as $\alpha + j\beta$, where α is the *attenuation constant* and β is the *phase constant* for the medium. Any electromagnetic field propagating through the medium can be described in terms of η and σ, hence the adjective "intrinsic."

As an electromagnetic wave propagates x meters through the medium, it will be attenuated by $\exp(\alpha x)$ and phase-shifted by βx radians. The electric fields (E-fields) and magnetic fields (H-fields) also exchange energy, so Z_w, the wave impedance, approaches η as the distance from the source increases (Figure H-1). Good insulators have $\rho \approx \infty$, so $\eta \approx (\mu/\epsilon)^{1/2}$ and $\sigma \approx j2\pi f(\mu\epsilon)^{1/2}$—electromagnetic waves travel through insulators with no energy loss. Good conductors have $\rho \approx 0$, so $\eta \approx (\pi f\mu\rho)^{1/2} + j(\pi f\mu\rho)^{1/2}$ and $\sigma \approx (\pi f\mu/\rho)^{1/2} + j(\pi f\mu/\rho)^{1/2}$—electromagnetic waves suffer substantial energy loss as they travel through conductors. The *skindepth* is defined as the distance by which a wave is attenuated by $\exp(1)$, so skindepth $\delta = 1/\alpha = [\rho/(\pi f\mu)]^{1/2}$ meters.

Figure H-2 shows a short dipole (length $l < \lambda/16$) immersed in an insulator with permeability μ and permittivity ϵ. The dipole carries a sinusoidal current I with wavelength λ meters. At a point (r, φ, θ) the peak electromagnetic fields are

$$E_r = \frac{2\pi \left(\frac{\mu}{\epsilon} \right)^{1/2} Il}{\lambda^2} \left[\left(\frac{\lambda}{2\pi r} \right)^2 - j \left(\frac{\lambda}{2\pi r} \right)^3 \right] \cos\theta \ \text{V/m}$$

$$E_\theta = \frac{\pi \left(\frac{\mu}{\epsilon} \right)^{1/2} Il}{\lambda^2} \left[j \left(\frac{\lambda}{2\pi r} \right) + \left(\frac{\lambda}{2\pi r} \right)^2 - j \left(\frac{\lambda}{2\pi r} \right)^3 \right] \sin\theta \ \text{V/m}$$

$$E_\varphi = 0 \ \text{V/m}$$

$$H_r = 0 \ \text{A/m}$$

$$H_\theta = 0 \ \text{A/m}$$

$$H_\varphi = \frac{\pi Il}{\lambda^2} \left[j \left(\frac{\lambda}{2\pi r} \right) + \left(\frac{\lambda}{2\pi r} \right)^2 \right] \sin\theta \ \text{A/m}$$

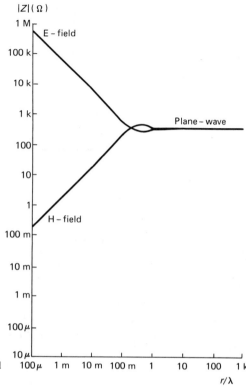

Figure H-1   Wave impedance of electric and magnetic fields

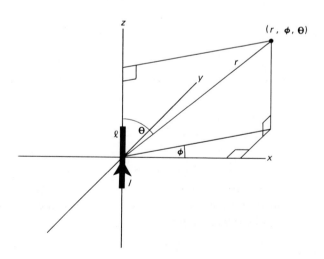

Figure H-2   Coordinate system for a small dipole

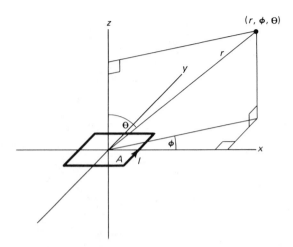

Figure H-3 Coordinate system for a small loop

In the near-field, $r < \lambda/2\pi$ meters, the E-field is greatest on the z-axis ($\theta = 0°$, $\pm 180°$) and the H-field is greatest on the xy-plane ($\theta = \pm 90°$). On the xy-plane this wave has wave impedance (Figure H-1)

$$Z_w = \left(\frac{\mu}{\epsilon}\right)^{1/2} \frac{j\left(\frac{\lambda}{2\pi r}\right) + \left(\frac{\lambda}{2\pi r}\right)^2 - j\left(\frac{\lambda}{2\pi r}\right)^3}{j\left(\frac{\lambda}{2\pi r}\right) + \left(\frac{\lambda}{2\pi r}\right)^2} \text{ ohms;}$$

$$|Z_w| \approx \left(\frac{\mu}{\epsilon}\right)^{1/2} \text{ ohms} \quad \text{for } r \gg \frac{\lambda}{2\pi} \text{ meters, and}$$

$$|Z_w| \approx \left(\frac{\mu}{\epsilon}\right)^{1/2} \left(\frac{\lambda}{2\pi r}\right) \text{ ohms} \quad \text{for } r \ll \frac{\lambda}{2\pi} \text{ meters.}$$

Figure H-3 shows a small loop of area A (perimeter $l < \lambda/2$ meters) immersed in an insulator with permeability μ and permittivity ϵ. The loop carries a sinusoidal current I with wavelength λ meters. At a point (r, φ, θ) the peak electromagnetic fields are

$$E_r = 0 \text{ V/m}$$

$$E_\theta = 0 \text{ V/m}$$

$$E_\varphi = \frac{2\pi^2 \left(\frac{\mu}{\epsilon}\right)^{1/2} AI}{\lambda^3} \left[\left(\frac{\lambda}{2\pi r}\right) - j\left(\frac{\lambda}{2\pi r}\right)^2\right] \sin\theta \ \text{V/m}$$

$$H_r = \frac{4\pi^2 AI}{\lambda^3} \left[j\left(\frac{\lambda}{2\pi r}\right)^2 + \left(\frac{\lambda}{2\pi r}\right)^3\right] \cos\theta \ \text{A/m}$$

$$H_\theta = -\frac{2\pi^2 AI}{\lambda^3} \left[\left(\frac{\lambda}{2\pi r}\right) - j\left(\frac{\lambda}{2\pi r}\right)^2 - \left(\frac{\lambda}{2\pi r}\right)^3\right] \sin\theta \ \text{A/m}$$

$$H_\varphi = 0 \ \text{A/m}$$

In the near-field, $r < \lambda/2\pi$ meters, the E-field is greatest on the xy-plane ($\theta = \pm 90°$), and the H-field is greatest on the z-axis ($\theta = 0°, \pm 180°$). On the xy-plane this wave has wave impedance (Figure H-1)

$$Z_w = \left(\frac{\mu}{\epsilon}\right)^{1/2} \frac{-\left(\frac{\lambda}{2\pi r}\right) + j\left(\frac{\lambda}{2\pi r}\right)^2}{\left(\frac{\lambda}{2\pi r}\right) - j\left(\frac{\lambda}{2\pi r}\right)^2 - \left(\frac{\lambda}{2\pi r}\right)^3} \ \text{ohms;}$$

$$|Z_w| \approx \left(\frac{\mu}{\epsilon}\right)^{1/2} \ \text{ohms} \quad \text{for } r \gg \frac{\lambda}{2\pi} \ \text{meters, and}$$

$$|Z_w| \approx \left(\frac{\mu}{\epsilon}\right)^{1/2} \left(\frac{2\pi r}{\lambda}\right) \ \text{ohms for } r \ll \frac{\lambda}{2\pi} \ \text{meters.}$$

These formulas are based on dipoles having infinite impedance and loops having zero impedance. But real circuits have finite, nonzero impedances. If a high-impedance circuit has impedance $Z = V/I > 376.7 \ \Omega$, it produces an E-field with wave impedance $Z_w = Z \ \Omega$ for $r \leqslant 376.7\lambda/ 2\pi Z$ meters. For $r > 376.7\lambda/2\pi Z$ meters the wave impedance follows the short-dipole curve (Figure H-4). Similarly, a low-impedance circuit with impedance $Z = V/I < 376.7 \ \Omega$ produces an H-field with wave impedance $Z_w = Z \ \Omega$ for $r \leqslant 376.7\lambda Z/2\pi$ meters. For $r > 376.7\lambda Z/2\pi$ meters the wave impedance follows the small-loop curve (Figure H-4). This reduction in the range of the wave impedance may be important when we are trying to calculate a shield's effectiveness for a circuit.

One final disturbing factor is standing waves. When the length of a conductor exceeds $\approx\lambda/4$ meters, the current in the conductor becomes nonuniform, thus reducing the emitted electromagnetic fields and reducing the voltages/currents induced by incident fields. The short-dipole and small-loop models are conservative and may indicate noise problems that really won't occur in practice.

In the "far-field," $r \geq \lambda/2\pi$ meters, the electric field and the magnetic field are roughly in balance, with wave impedance $Z_w \approx \eta$ ($\approx 376.7 \ \Omega$ in air). In the "near-field," $r < \lambda/2\pi$ meters, Z_w can be greater than, equal to, or less than η. If $Z_w > \eta$, the electric field dominates and tries to induce voltage differences in nearby conductors. If $Z_w < \eta$, the magnetic field dominates and tries to induce current flow in nearby conductors. A circuit's impedance, and the impedance of an electromagnetic wave hitting the circuit, determine how much noise is injected into the circuit. We have the greatest noise pickup when a high-impedance wave hits a high-impedance circuit (capacitive crosstalk) and when a low-impedance wave hits a low-impedance circuit (inductive crosstalk).

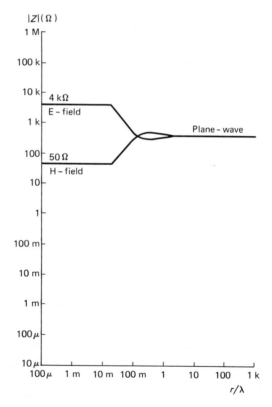

Figure H-4 Impedance of electric and magnetic fields with finite source impedance

RECOMMENDED READING

SCHELKUNOFF, S. A., *Electromagnetic Waves*. Princeton, NJ: D. Van Nostrand
   Co., Inc., 1943.

APPENDIX I: COMPLEX ARITHMETIC

Let V and I be two complex numbers (or vectors, Figure I-1)

$$V = a + jb,$$
$$I = c + jd,$$

where j is the square root of -1. The magnitudes of V and I are

$$|V| = (a^2 + b^2)^{1/2},$$
$$|I| = (c^2 + d^2)^{1/2},$$

and the phase angles of V and I are

$$\theta = \arctan \frac{b}{a},$$

$$\varphi = \arctan \frac{d}{c},$$

letting us write

$$V = |V| \cos \theta + j|V| \sin \theta,$$
$$I = |I| \cos \varphi + j|I| \sin \varphi.$$

The sum of V and I is (Figure I-2)

$$V + I = (a + c) + j(b + d).$$

The difference between V and I is

$$V - I = (a - c) + j(b - d).$$

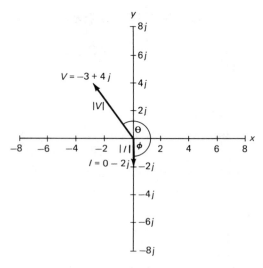

Figure I-1   Vectors on the complex plane

The product of V and I is

$$VI = (ac - bd) + j(ad + bc)$$
$$= |V||I| \cos(\theta + \varphi) + j|V||I| \sin(\theta + \varphi).$$

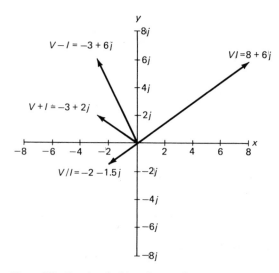

Figure I-2   Results of arithmetic operations on the vectors

The ratio of V to I is

$$\frac{V}{I} = \frac{ac + bd}{c^2 + d^2} + j\,\frac{bc - ad}{c^2 + d^2}.$$

$$= \frac{|V|}{|I|}\cos(\theta - \varphi) + j\,\frac{|V|}{|I|}\sin(\theta - \varphi).$$

ANNOTATED BIBLIOGRAPHY

Consumer Electronics Systems Technician Interference Handbook—Audio Rectification. Washington, D.C.: Consumer Electronics Group/Electronic Industries Association, no date. Describes methods for diagnosing and fixing audio rectification problems in audio equipment.

Consumer Electronics Systems Technician Interference Handbook—TV Interference. Washington, D.C.: Consumer Electronics Group/Electronic Industries Association, no date. Describes methods for diagnosing and fixing television interference (TVI) problems.

Design Techniques for Interference-Free Operation of Airborne Electronic Equipment. Springfield, VA: NTIS (AD 491 988), 1952. Covers shielding and filtering practices in great detail. Also has good sections on grounding, bonding, how to EMI-proof motors and switches, and how to fix EMI problems built into electronic equipment.

Interference Reduction Guide for Design Engineers, Vol. 1. Springfield, VA: NTIS (AD 619 666), 1964. Good sections on the frequency spectrum of pulses, interference generators, grounding, bonding, cabling, and shielding. Emphasizes design for RFI control and contains many useful graphs and tables of design data.

BLAKESLEE, THOMAS R., *Digital Design with Standard MSI and LSI*. New York: John Wiley & Sons, 1975. Two excellent chapters cover "Nasty Realities" encountered in digital systems, including problems caused by wires.

BLOOD, WILLIAM R., JR., *MECL System Design Handbook*, 4th ed. Phoenix, AZ: Motorola Semiconductor Products, 1983. Covers all aspects of designing high-speed digital systems, with several chapters on transmission lines and cables.

DENNY, H. W., et al., *Grounding, Bonding, and Shielding Practices and Procedures for Electronic Equipments and Facilities*, Vol. I. Springfield, VA: NTIS (AD A022 332), 1975. Covers the grounding of buildings and equipment for lightning protection, power-fault protection, and reference grounds. Also covers bonding, shielding, noise-coupling paths, and the protection of equipment from electromagnetic pulses.

DENNY, H. W., et al., *Grounding, Bonding, and Shielding Practices and Procedures for Electronic Equipments and Facilities*, Vol. II. Springfield, VA: NTIS (AD A022 608), 1975. Covers construction, inspection, and maintenance procedures for ground systems, bonds, and shields.

DENNY, H. W., et al., *Grounding, Bonding, and Shielding Practices and Procedures for Electronic Equipments and Facilities*, Vol. III. Springfield, VA: NTIS (AD A022 871), 1975. Covers cost estimating for grounding systems.

DENNY, HUGH W., *Grounding for the Control of EMI*. Gainesville, VA: Don White Consultants, Inc., 1983. Covers the design of grounding systems to minimize noise problems within and between pieces of equipment.

EVERETT, WOODROW W., JR., *Topics in Intersystem Electromagnetic Compatibility*. New York: Holt, Rinehart and Winston, Inc., 1972. Has some excellent chapters on the high-frequency characteristics of components, grounding, bonding, and shielding.

FICCHI, ROCCO F., ed., *Practical Design for Electromagnetic Compatibility*. New York: Hayden Book Co., Inc., 1971. Covers interference sources, component characteristics, shielding, grounding, and bonding.

GROVER, FREDERICK W., *Inductance Calculations*. Instrument Society of America, 1973. Contains formulas and tables for calculating the inductance of wires, transmission lines, and coils.

GUNSTON, M. A. R., *Microwave Transmission-Line Impedance Data*. New York: Van Nostrand Reinhold Co., 1972. Contains a wealth of data on designing transmission lines.

HILBERG, WOLFGANG, *Electrical Characteristics of Transmission Lines*. Dedham, MA: Artech House Books, 1979. Derives the characteristic equations for many types of transmission lines using conformal mapping.

KEENAN, R. KENNETH, *Decoupling and Layout of Digital Printed Circuits*. Pinellas Park, FL: TKC, 1985. Covers the design of power/ground bussing on printed circuit boards in great detail, including how to select components for maximum effectiveness.

KEENAN, R. KENNETH, *Digital Design for Interference Specifications*. Pinellas Park, FL: TKC, 1983. Covers the design of digital systems to minimize EMC problems, including board and cable design, power/ground bussing, power supply design, and shielding. Tells how to set up a small EMC lab—for $1500 to $5000—that can perform developmental tests on new products.

MARDIGUIAN, MICHEL, *How to Control Electrical Noise*. Gainesville, VA: Don White Consultants, Inc., 1983. Tells how to design products to meet EMC requirements. Covers EMI sources, shielding, printed circuit board and cable design, grounding, bonding, and EMC testing.

MARDIGUIAN, MICHEL, *Interference Control in Computers and Microprocessor-Based Equipment*. Gainesville, VA: Don White Consultants, Inc., 1984. Tells how to design microprocessor-based equipment to meet EMC requirements. Covers printed circuit board, cable, and power supply design, grounding, shielding, filtering, and EMC testing.

NELSON, WILLIAM R., *Interference Handbook*. Wilton, CT: Radio Publications, Inc., 1984. Covers radio-frequency interference (RFI) that affects radios, televisions, and audio equipment. Full of information on locating RFI sources and reducing RFI emissions and RFI susceptibility.

OTT, HENRY W., *Noise Reduction Techniques in Electronic Systems*. New York:

John Wiley & Sons, 1976. A very readable book covering every aspect of electronic system design. Emphasizes the design process, and contains a wealth of practical information on how to design electronic systems that do what they are supposed to do. Highly recommended.

SCHELKUNOFF, S. A., *Electromagnetic Waves*. Princeton, NJ: D. Van Nostrand Co., Inc., 1943. This book is the "bible" of electromagnetic field theory and shielding theory.

WHITE, DONALD R. J., *EMI Control in the Design of Printed Circuit Boards and Backplanes*. Gainesville, VA: Don White Consultants, Inc., 1981. Focuses on the design of printed circuit boards to meet EMC requirements.

WHITE, DONALD R. J., *A Handbook on Electromagnetic Shielding Materials and Performance,* 2nd ed. Gainesville, VA: Don White Consultants, Inc., 1980. Covers shielding theory and the choice of materials for shielding.

WHITE, DONALD R. J., *A Handbook Series on Electromagnetic Interference and Compatibility,* Vol. 3. Gainesville, VA: Don White Consultants, 1973. This book discusses many methods for eliminating problems uncovered during EMC testing, including component selection, grounding, bonding, shielding, and filtering.

WHITE, DONALD R. J., *Shielding Design Methodology and Procedures*. Gainesville, VA: Interference Control Technologies, 1986. Takes you through the design of shields for a system in a step-by-step process, complete with worksheets and design data.

GLOSSARY

AC. Alternating current.

Backplane. Printed circuit board used to interconnect other circuit boards.

Backporching. Rise or fall of a signal in a series of steps, caused by improper termination of a transmission line.

Balanced line. Transmission line designed and built to provide equal resistance, inductance, and capacitance on the signal line and the signal-return line. Twisted-pair lines are balanced, coaxial cables are unbalanced.

Clock duty cycle. Ratio of a clock signal's active period to its total period.

CMOS IC. Complementary metal-oxide semiconductor integrated circuit, uses both N-channel and P-channel field-effect transistors.

Common-mode signal. Signal applied to a signal line and its signal-return line equally.

DC. Direct current.

Differential driver. Circuit that sinks approximately as much current from the signal-return line as it sources onto the signal line. Used to drive differential receivers through balanced lines.

Differential-mode. Signal applied between a signal line and its signal-return line.

Differential receiver. Circuit that detects the voltage between a signal line and its signal-return line, while ignoring the voltage common to both lines.

ECL IC. Emitter-coupled logic integrated circuit, uses bipolar transistors.

Electromagnetic compatibility (EMC). An electronic system's ability to operate with other electronic systems without degrading the performance of any of the systems.

Electromagnetic interference (EMI). Performance degradation of an electronic system due to electromagnetic fields.

Electromagnetic susceptibility (EMS). An electronic system's ability to tolerate electromagnetic fields without performance degradation.

EMC. Electromagnetic compatibility.

EMI. Electromagnetic interference.

EMS. Electromagnetic susceptibility.

End-of-life tolerance. Difference between a component's true value at the end of its useful life and its nominal value, expressed as a percentage of the nominal value.

EOL tolerance. End-of-life tolerance.

Ferrite bead. Cylinder with one or more holes, made from a ferrimagnetic material with high resistivity. When slipped onto a wire, a ferrite bead acts like a short circuit at low frequencies and like a resistor at high frequencies.

FET.   Field-effect transistor.

Glitch.   An unwanted transient voltage spike on a signal.

Groundplane.   Low-impedance conductive surface that serves as both a signal reference and a current return.

IC.   Integrated circuit.

Input-data strobe.   Signal used to load unsynchronized data into a flipflop or latch for processing by a synchronous circuit.

MOS IC.   Metal-oxide semiconductor integrated circuit, uses field-effect transistors (includes CMOS, NMOS, and PMOS).

Motherboard.   Printed circuit board with logic circuits that is also used to interconnect other circuit boards.

NMOS IC.   N-type metal-oxide semiconductor integrated circuit, contains N-channel field-effect transistors.

Noise margin.   Voltage difference between a driver's output voltage and the minimum input voltage required by a receiver.

PCB.   Printed circuit board.

Permeability.   Magnetic flux density in a material divided by the magnetizing field. Air and vacuum have a permeability of 1.25663706 microhenries per meter.

Permittivity.   Electric flux density in a material divided by the electric field. Air and vacuum have a permittivity of 8.85418782 picofarads per meter.

Plane wave.   An electromagnetic field far away from the source, with impedance $Z_w = (\mu/\epsilon)^{1/2} \approx 376.7\Omega$ in air or vacuum.

PMOS IC.   P-type metal-oxide semiconductor integrated circuit, contains P-channel field-effect transistors.

Reference.   A conductor that is supposed to hold two or more points at the same potential.

Resistivity.   Conduction current density in a material divided by the electric field. Electrical-grade copper (100% IACS) has a resistivity of 17.2 nano-ohm-meters.

Return.   Conductor that is supposed to carry current from the load back to the source.

RF.   Radio-frequency, approximately 300 kHz and up.

RFI.   Radio-frequency interference. See *Electromagnetic interference*.

Skindepth.   Depth at which an electromagnetic field is attenuated to $1/e$ of its value at the surface of a conductor.

Skin effect.   Concentration of alternating current near the surface of a conductor.

Termination Network. Circuit placed at the end of a long transmission line to reduce signal reflections.

TTL IC. Transistor-transistor logic integrated circuit, uses bipolar transistors.

INDEX